Progressing Cavity Pumps, Downhole Pumps, and Mudmotors

Progressing Cavity Pumps, Downhole Pumps, and Mudmotors

Dr. Lev Nelik, P.E., APICS
and Jim Brennan

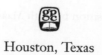

Houston, Texas

Gulf Publishing Company
2 Greenway Plaza, Suite 1020
Houston, TX 77046

10 9 8 7 6 5 4 3 2 1

Printed and bound by CPI Group (UK) Ltd, Croydon, CR0 4YY

Transferred to digital print 2012

Text design and composition by Ruth Maassen.

ISBN 0-9765113-1-2

Contents

About the Authors

 Dr. Nelik has 30 years of experience with pumps and pumping equipment. He is a registered professional engineer who has published over fifty documents on pumps and related equipment worldwide, including a "Pumps" section for the *Encyclopedia of Chemical Technology* (John Wiley & Sons, Inc.), a section for the *Handbook of Fluids Dynamics* (CRC Press), and his book *Centrifugal and Rotary Pumps: Fundamentals with Applications* (CRC Press). He is a president of Pumping Machinery, LLC, specializing in pump consulting, training, and equipment troubleshooting. His experience in engineering, manufacturing, sales, and management includes Ingersoll-Rand (engineering), Goulds Pumps (technology), Roper Pump (vice president of engineering and repair/overhaul), and Liquiflo Equipment (southeast regional sales manager, and later a president). Dr. Nelik is an advisory

committee member for Texas A&M University's International Pump Users Symposium, an editor of *Pump Magazine*, an advisory board member of *Water and Wastes Digest*, an editorial advisory board member of *Pumps & Systems* magazine, and a former associate technical editor of *Journal of Fluids Engineering*. He is a full member of the American Society of Mechanical Engineers (ASME), and is certified by the American Production and Inventory Control Society (APICS). He is a graduate of Lehigh University with a Ph.D. in mechanical engineering and a master's in manufacturing systems. He teaches pump training courses in the United States and worldwide, and consults on pump operations, engineering aspects of centrifugal and positive displacement pumps, and maintenance methods to improve reliability, obtain energy savings through improved efficiency, and optimize pump-to-system operation.

Jim Brennan has over 35 years of experience with pumps and pumping systems. He is a member of the Society of Petroleum Engineers and holds a bachelor of science from Drexel University in mechanical industrial engineering. His experience includes work with single-, two-, and three-screw pumps; gear pumps and vane pumps; and power generation, lubrication, gas sealing, fluid power, and marine and oil-field pipeline applications. He is currently a special projects manager at Imo Pump, a member of the Colfax Pump Group. His previous assignments at Imo Pump included engineering manager supervising the quality assurance function, field service, and repair business, as well as pump design, development, testing, and applications. He has authored many papers and articles and has spoken at a number of industry conferences on pumps and pump-related subjects. He has also taught rotary positive displacement pump operation.

PREFACE

This publication describes the fundamentals of progressing cavity (PC) pumps. Although formally classified by the Hydraulic Institute as a *single-screw* type, a subset (group) of the rotary class pumps, they are better known as PC pumps and are used in a wide variety of applications. They are especially suited for tough, nasty, multiphase fluids with gas and solids in suspension, at relatively moderate temperatures (under 350°F).

When designed to operate in reverse (i.e., to cause shaft rotation by the supplied differential pressure), these pumps can essentially operate as hydraulic drives. In the oil-drilling and exploration industry, such motors are called *downhole mud-motors* (DHMs), referring to drilling "mud" that provides hydraulic medium to drive the rotor, lubricate the drill bit, and remove the debris.

In oil production, these types of pumps are referred to as *downhole pumps* (DHPs), operating at depths of several thousand feet, pumping oil (usually mixed with other fluids, gases, and sand) to the surface.

Presently available publications on this subject are scarce. Technical papers appear mainly in specialized symposiums, such as oil exploration conferences, and usually deal with a specific subject or aspect of applications, with little coverage of the pump or mudmotor design itself, especially its hydraulic sections. Textbooks are also rare, and mostly concentrate on overview rather than the design details or operating principles. The rather complex internal geometry of these units thus remains vaguely understood, even within the engineering community; it has not been explored sufficiently, perhaps in fear of complicated mathematics and three-dimensional surfaces. However, the complexity of these PC machines can be understood, and their mysteries unveiled, by a step-by-step explanation of their geometry, uninhibited by complex math. Instead, we will use simple (or simplified) formulas that show the geometry of a pump, including the "heart of the issue"—a "progressing" cavity, as it moves fluid from inlet to discharge. This explanation will start with a (simpler) single-lobe rotor geometry and will gradually expand to multilobe units.

The formal approach to analyzing and understanding these types of pumps and mudmotors is just beginning to emerge, but it is still more art than science. There is no universally accepted nomenclature to denote major and minor diameters, eccentricity, rotor/stator fits, or thermal shrink rates. The nomenclature used in this book is a first attempt to unify and standardize such symbols, with the goal of eventually being incorporated into the ISO standard to benefit publications and to facilitate communication within the technical community.

Hopefully, this publication will serve as a timely teaching and reference tool for practicing engineers, designers, and pump and mudmotor users for better understanding and appreciation of the fundamentals of this pump type.

Acknowledgments

I am grateful to the staff of Gulf Publishing Company for their editorial work, comments, and suggestions.

I would like to extend my thanks to many organizations and individuals whose valuable input was helpful. In my 30 years in the pump business, it would be impossible to mention all of my "pumping comrades" whose influence, help, discussions, and support have ultimately made it possible for me to author my publications. But their help and professionalism are within these pages, and I am grateful for their presence in my life.

Particular appreciation goes to Monoflo Pump Company (currently a division of National Oilwell Company), for their valuable assistance with application examples and illustrations.

My loving appreciation goes to my wife, Helaine, for her support and understanding of my various endeavors related to pumps, and for allowing me to occasionally turn the basement into a laboratory inaccessible for other (more important) household needs. To our children, Adam, Asher, and Joshua, wishing them happiness and success in their personal and professional lives.

Dr. Lev Nelik, P.E., APICS
Pumping Machinery, LLC
Atlanta, GA
www.PumpingMachinery.com

When Lev asked me to co-author this book, I was flattered as well as awed by the amount of material he had already created. We have worked together for many years, presenting the short course "Positive Displacement Pumps" at Texas A&M University's International Pump Users Symposium. His energy level and enthusiasm rubs off on students as well as co-presenters. His work is truly the "meat and potatoes" of progressing cavity design and operating theory. I wish to thank him as well as my employer, Colfax Pump Group and its many associates, for supplying me opportunities to learn about pumps and pump systems and to share some of that knowledge with others.

Jim Brennan
Imo Pump, Colfax Pump Group
Monroe, NC

Nomenclature and Abbreviations

c_m	radial fit on minor diameter (in.)
c_j	radial fit on major diameter (in.)
d, D	diameters of construction circles; d_r – rotor, D_s – stator (in.)
d_m	rotor minor diameter (in.)
d_j	rotor major diameter (in.)
D_m	stator minor diameter (in.)
D_j	stator major diameter (in.)
e	eccentricity (distance between rotor and stator centers; also a radius of the generating circle)
FHP	fluid horsepower (hp)
BHP	break horsepower (hp)
PCP	surface-mounted progressing cavity pump
DHP	downhole pump
DHM	downhole mudmotor
lobe ratio	$N_r{:}N_s$
NPSHa	net positive suction head, available (ft)

NPSHr	net positive suction head, required (ft)
N_r	number of rotor lobes
N_s	number of stator lobes
P_r	rotor pitch (in.)
P_s	stator pitch (in.)
rpm	input shaft rotational speed (rev/min)
rpm_n	nutational speed of rotor center around stator center (rev/min) = rpm $\times N_r$
Q	total flow, at operating load (gal/min = gpm)
Q_o	flow through working passages (not the slip passage), $Q = Q_o$ at zero load, when $\Delta p = 0$
Q_{slip}	slip (leakage) (gal/min = gpm)
q	unit flow, at operating load (gal/rev = gpr)
q_o	unit flow at zero load (gal/rev = gpr)
π	3.14 (pi)
V	volume (cavity) (in.^{3}); also velocity (ft/sec)
A_{fluid}	net opening at the cross section between stator void and rotor metal (in.^{3})
t_{mo}	elastomer thickness on stator minor diameter prior to cooling (in.)
t_{jo}	elastomer thickness on stator major diameter prior to cooling (in.)
t_m	elastomer final (measured) thickness on stator minor diameter (in.)
t_j	elastomer final (measured) thickness on stator major diameter (in.)
Δ_m	absolute shrinkage on minor diameter (in.)
Δ_j	absolute shrinkage on major diameter (in.)
S_m	relative shrinkage on minor diameter [in./(in. $\times$ °F)]

S_j	relative shrinkage on major diameter [in./(in. × °F)]
T	torque at operating conditions (ft × lbs)
p_d	discharge (exit or outlet) pressure (psig)
p_s	suction (inlet) pressure (psig)
Δp	differential pressure (psi)
ε_m	thermal coefficient of expansion on minor diameter [in./(in. × °F)]
ε_j	thermal coefficient of expansion on major diameter [in./(in. × °F)]
η_{vol}	volumetric efficiency
η	total (net) efficiency
MTBF	mean time between failures

INTRODUCTION

Progressing cavity surface-mounted pumps (PCPs), downhole pumps (DHPs), and downhole mudmotors (DHMs) have received renewed attention from the pumping community in recent decades, although the original design dates back to the early 1930s, when Dr. Rene Moineau invented this pump type and applied it as a supercharger for an airplane engine. Its ability to pump multiphase fluids containing liquid, gas, air, solids, or vapor often makes it a "last resort" choice, when other types of pumps cannot be applied. This type of pump belongs to a subclass of rotary pumps and is officially included in the group "screw pumps" within the Hydraulic Institute standards (Ref. 1), as shown in Fig. 1(a,b).

Despite being classified by the Hydraulic Institute as a screw-type pump, users and manufacturers (especially in North America) usually call this a progressing cavity (PC) pump, reserving the category "screw pumps" for multiple-screw types (i.e., two- [sometimes called twin-screw pumps] and three-screw pumps). In fact, there is much more similarity between

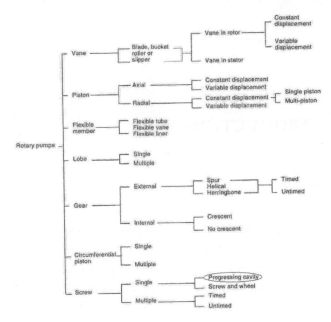

FIGURE I(A)
Types of pumps (part of the diagram from Hydraulic Institute Standards, Ref. 1).

applications of *multiple*-screw pumps (two- and three-screw) than *single*-screw (i.e., progressing cavity). Progressing cavity pumps are often applied to "nasty," difficult applications such as wastewater treatment, solids in suspension, highly abrasive slurries, etc., while two- and three-screw pumps are mostly used for cleaner fluids such as in oil and fuel transport. There are occasional exceptions and application boundaries are crossed at times. However, in Europe (especially in Russia), PC pumps have retained their Hydraulic Institute classification

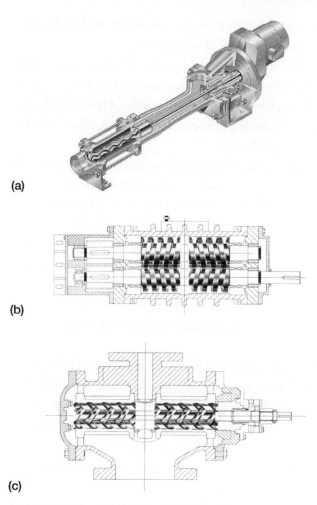

(a)

(b)

(c)

FIGURE 1(B)
Similarities between the families of positive displacement pumps:
(a) Single-screw pump (progressing cavity pump) (courtesy Monoflo).
(b) Two-screw pump (courtesy Colfax Pump Group).
(c) Three-screw pump (courtesy Colfax Pump Group).

and are usually referred to as screw pumps, both in Russian publications and in English translations.

Overall, the term *progressing cavity pumps* is more widely accepted, and we will use this terminology (PC pumps), given the above explanation, in this publication. Hydraulically, the principle of a moving "progressing" cavity applies to PCPs, DHPs, and DHMs alike.

The reasons for the relative obscurity of PC pumps are still unclear, and pump users who have researched, understood, and installed these machines often wonder why they did not become aware of their great benefits until relatively recently. The many benefits of PC pumps are only beginning to become known in the North American market, as well as internationally. The benefits of PC pumps far outweigh their limitations, and the purpose of this book was to present these benefits, as well as drawbacks, and to relate them to their unique design features and capabilities.

The great majority of all pumps in the world belong to a class called *centrifugal* pumps. The other category, the *positive displacement* (PD) type, represents a smaller proportion of the total pump population. Although various studies disagree somewhat, it is probably safe to estimate positive displacement designs at about one-third of the world pump population, and centrifugal at two-thirds.

From an historical perspective, the earliest applications of pumps were simple—to pump water. Centrifugal pumps were the ideal candidates for that job because water has low viscosity and lubricity, resulting in low friction losses inside the hydraulic passages of centrifugal pumps. A significant part of centrifugal pump engineering work in the early twentieth century focused on optimizing the hydraulics of centrifugal pumps. Numerous papers and books have been written on impeller designs, volute configurations, diffusor vanes, etc. Centrifugal pumps vary in size, from tiny fractional horsepower for residential home needs, to over 60,000 bhp boiler feed pumps for steam turbine power stations.

When chemical plants emerged in the early 1900s, the chemicals to be pumped were also low-viscosity, waterlike substances. Therefore, PD pumps were not developed until large steam-generating stations began to emerge, which brought in large reciprocating steam piston pumps.

The class of PD pumps is composed of the reciprocating and rotary types. These types are illustrated later in this book, and are formally categorized according to the Hydraulic Institute classification chart. Both reciprocating and rotary pumps push fluid from an inlet toward discharge ports (i.e., they generate flow, as compared to centrifugals, which are said to generate pressure). However, reciprocating PD pumps are generally larger and their mechanical linkages encounter significant stresses due to acceleration, stoppage, and reacceleration of the piston. Rotary pumps were developed as a compromise between the large, slow-running reciprocating pumps, and the smaller, faster-operating centrifugals. Progressing cavity pumps are rotary pumps and, as such, are cousins to gear and lobe pumps, although they more closely resemble multiple-screw pumps.

Although they were originally conceived by Dr. Rene Moineau shortly after WWI, for the above reasons these pumps did not become popular until the last 15–20 years. Dr. Moineau, a brilliant engineer, had the idea to insert a single-lead rotor into a double-lead stator, thus creating a cavity between the two. Rotary motion of the rotor displaces the cavity, which is the main concept of the "progressing" cavity. Nevertheless, at the beginning, his design could not effectively compete against the smaller and less-expensive centrifugal pumps. For water, the centrifugal pump would prevail against the PC design for many years.

As various industries began to mature, especially after WWII, new chemicals were invented and were produced at exponential rates. Many of these chemicals were viscous materials, significantly different from low-viscosity water. For example, the paper industry produced new requirements for transferring pulp; wastewater treatment plants began to grow in size and production volumes in response to population growth; and increased consumer and industrial production resulted in greater turnaround demands.

At the same time, manufacturing technologies made significant breakthroughs in machinery designs. Five-axis machines, equipped with the latest electronic controllers, made it much easier and less expensive to cut the complex three-dimensional shapes of PC rotors and stators. In short, by about the mid-1980s, PC pumps began competing for recognition.

The designs of the multilead units have now arrived, and applications of surface-mounted pumps have expanded to include downhole pumps for oil production, as well as "pumps-in-reverse"(called mudmotors) for the drilling and oil exploration industry.

Chapter 1

BENEFITS OF PROGRESSING CAVITY PUMPS

From the standpoint of flow and pressure range, PC pumps compete well with any other type of pump design. They can handle pump flows from fractions of gallons per minute to several thousand gallons per minute. Pressure capability depends on the number of stages (leads of the stator), and typically reaches 800–1,000 psi. The range of fluids they can handle is enormous, with viscosities ranging from waterlike (1 cSt), to fluidlike (clay, cement, and sludge with viscosities up to 1,000,000 cSt).

Since the rotor and stator have an interference fit (a plated metallic rotor in an elastomer-lined stator), and low rotating speed, the internal shear rates are very low. When applied in the food industry, these pumps are sometimes known to pump cherries or apples, which move through the internal passages with no damage. The pulsation-free flow and quiet operation of PC pumps are additional advantages for shear-sensitive pumpages.

1

PC pumps are excellent self-primers and have good suction characteristics. They are tolerant to entrained air and gases, and produce minimal churning or foaming.

Another major feature of these machines is their high tolerance to contamination and abrasion. Often called a "last-resource pump," PC pumps are often utilized for extremely abrasive applications, because of a unique property of the elastomer that lines up the stator tube and resists abrasion. These elastomers are made from regular rubber (Buna) or more exotic materials, such as Viton®, Teflon®, and others.

As versatile as they are, PC pumps do have limitations, primarily size. To prevent flow "slip" (leakage from higher discharge pressure back to suction), the number of rotor/stator leads (stages) must be increased as pressure is increased. This, in turn, increases the overall length of the unit. For such high-pressure applications, it is sometimes difficult to retrofit an existing installation, where smaller pumps (such as centrifugal) have been operating in the past. However, when space is not an issue, this limitation is not a factor.

Another reason for PC pumps' larger size is their low speed, which requires a gear reducer (or a belt drive) between the motor driver and the pump. This can result in added cost. However, the recent advent of variable frequency drives (VFDs) has allowed elimination of gear reducers and, at the same time, has introduced a new capability to vary flow while pumping against a given pressure.

Another PC pump limitation is fluid compatibility with the elastomer. Some chemicals may cause problems with the elas-

tomer and others may cause swelling. For corrosive applications, Viton® or even Teflon® stators are selected. Elastomers
also impose certain temperature limitations, as compared to
fully metallic pumps. Typically, PC pumps are used where temperatures do not exceed 300–350°F. PC pumps should not run
dry, except for a very short time, because heat generated at the
interference between the rotor and stator may cause elastomer
failures, often called "chunking" or "debonding."

Chapter 2

THREE MAIN TYPES OF PROGRESSING CAVITY MACHINES: PCPs, DHPs, AND DHMs

As shown in Figs. 2 and 3, a PC pump basically consists of a *hydraulic section* (a rotor inside a stator) and a *drive frame* transmitting the shaft rotation to the hydraulic section by means of a *connecting rod*. The connection is accomplished by various means: a pin joint, a universal joint (such as Cardan®), or a flexible joint arrangement.

A driver (motor, diesel, or other prime mover) is coupled to an input shaft, transmitting its energy into the mechanical energy of shaft rotation. The shaft is supported by bearings in the pump drive frame, and drive shaft rotation is translated into eccentric motion of a rotor in the hydraulic section. The rotor forms a cavity between itself and a stator, as will be explained in detail. The eccentric motion of the rotor displaces the fluid within the cavity, which "moves" (i.e., "progresses"—hence,

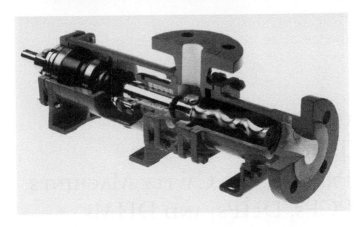

FIGURE 2
Typical progressing cavity pump (courtesy Monoflo Pump).

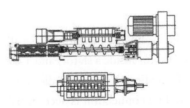

FIGURE 3
Progressing cavity cross section (courtesy Monoflo Pump).

the term) along the axis from suction (inlet) to discharge (exit) (i.e., pumping the fluid against the discharge pressure). Since the fluid is actually mechanically displaced, PC pumps belong to a subclass of rotary pumps.

If such a pump is connected to a generator instead of an electric motor or any other load (e.g., a drill bit), and the rotor is allowed to rotate driven by the differential pressure, the pump essentially becomes a *hydraulic motor*, per terminology of PD machines. (Note the similarity with centrifugal pumps operating as hydraulic recovery turbines when operated in reverse.) It utilizes the fluid's energy to convert it to the mechanical energy of the rotating shaft—from the eccentric rotor, via the connecting rod, to the drive frame shaft, and to the load. Thus, when mechanical energy is converted to hydraulic energy, the unit operates as a pump; conversely, when the hydraulic energy is converted to mechanical rotation, the unit operates as a hydraulic motor. In oil exploration drilling, such reverse operating units are called "mudmotors," which utilize "mud" as a working fluid to drive the rotor inside the stator, as well as to lubricate the drill bit and flush out the debris. In that case, the hydraulic section (rotor and stator) is called a "power section," denoting its utilization to transmit power to the drill bit. A complete drilling unit (a mudmotor) is shown in Fig. 4, and contains the power section itself, bearings, seals, drill bit, stabilizers, and other auxiliary equipment. Such a unit (referred to as a "string" by oil drillers) consists of a mudmotor, piping, and a drilling mechanism, all of which are inserted into and operate in a hole (i.e., future potential oil well). At the surface, the mud is pumped into the hole, usually by large, high-pressure triplex reciprocating pumps. A good overview of this can be found in Ref. 2.

The commercial logistics of the supplied equipment differ for mudmotors and PC pumps. Surface-mounted PC pumps are typically installed in facilities such as process and

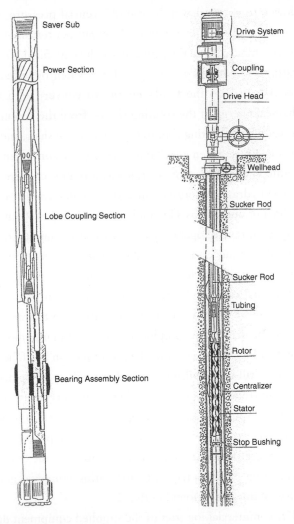

FIGURE 4
Complete units: A downhole mudmotor (left, DHM) (courtesy of Dresser Security Company), and a downhole pump (right, DHP) (courtesy of PCM Company).

petrochemical plants, paper mills, utilities, wastewater treatment plants, etc. They require much less space than the complete "string" of downhole mudmotor applications. A surface pump manufacturer supplies a complete pump unit, often mounted on a base plate and coupled to an electric motor or other driver. The customer or a contractor's millwright crew connects the piping and electric power, and the pump is ready to operate. In the case of mudmotors (DHMs), however, the power section is typically supplied by a subvendor (often the same one who manufactures complete PC pumps and has the facilities and know-how to produce power sections—cutting the rotors and injecting the elastomers into stator tubes). The subvendor often then supplies power sections to a mudmotor (complete unit) manufacturer, who cuts the required threads on the stator tube to connect to the rest of the mudmotor. These mudmotors are then sold (or, more often, rented) to the drillers who own the rigs and perform drilling contract jobs for the oil companies. Downhole pumps (see Fig. 4, DHP) are logistically handled according to the specifics of the industry in which they are applied, similar to DHMs. They are essentially rotor/stator sections, driven via a long drive shaft (in sections), or by a submersible motor.

Power sections (which are actually pumping sections for downhole pumps) are supplied unthreaded by the subvendors, for subsequent completion by the "string" manufacturer. This commercial practice is part of a tradition, history, and culture in the oil industry dating back many years.

PC pumps have a low "lobe ratio" (to be discussed in detail in Chapter 4), such as 1:2 (a one-lobed rotor inside of a two-lobed stator), although 2:3 lobe ratio designs have recently been introduced. Mudmotors usually have multilobe configurations to as high as 9:10. Geometry and space constraints inside the drilled hole require that power sections must be as short as possible (i.e., having fewer stages), with more pressure drop per stage (i.e., more power and torque concentration per unit length). Multilobe design satisfies such requirements, albeit with limitations and drawbacks that will be discussed in Chapter 4.

In the remainder of this book, we will concentrate on the details of the hydraulic section (rotor/stator) of pumps: surface-mounted, downhole (DHPs), and motors (DHMs), including their geometry, performance calculation and evaluation, and elastomer behavior. References are provided for more information on drilling applications, pumping conditions, and auxiliaries (seals, bearings, piping, etc.).

Chapter 3

OPERATING PRINCIPLES OF HYDRAULIC SECTIONS (ROTOR/STATOR PAIRS)

If one examines the geometry of a rotor, it will appear amazingly similar to a bolt or a screw, having thread crests with very smooth, blended, continuous transitions from one section to another along the axis. The same applies to a stator: it resembles a nut into which the rotor is inserted. Stator geometry is, likewise, smooth and continuously blended along the axis. What differentiates a rotor/stator pair from a regular mechanical screw/nut couple is that the number of lobes of the rotor does not equal that of the stator. A screw and matching nut have the same number of leads and the same pitch; otherwise, they would not go together. The number of leads (lobes) of a stator, on the other hand, is equal to the number of rotor leads *plus one* ($N_s = N_r + 1$). Their pitches are different, too. A *pitch* is an axial distance between the start of a particular lobe and its end (a 360-degree wraparound). This is different from a *lead*, which is an axial distance between two adjacent lobes.

For example, for a five-lobed stator, there are five leads per one pitch. You could not, therefore, insert a rotor into a stator and "thread it in," as you would for a screw and a nut. The profile geometries of a rotor and a stator are made differently from those of screws and nuts. The center of a screw always stays lined up with the center of a matching nut, but the center of a rotor changes its position along its axis. Each individual cross section of a rotor is offset from the center of a "matching" stator, from section to section. In addition, rotor sections "whirl around" the center of a stator during operation. This whirling motion of a rotor center around the center of a stator is called "nutation." This intentional lack of symmetry of a rotor in relation to a stator creates a complex cavity between them, and the nutation causes this cavity to be displaced (i.e., to progress) along the axis during the rotor rotation. Figure 5 illustrates this mechanism for a case of a 1:2 lobe unit, as further explained below.

For multilobe units, the same principle applies, except that there are *multiple* cavities created, which all move along the axis of the stator during the rotor revolution, and the total fluid displacement per one revolution of input shaft is equal to the product of the net (open) cross-sectional flow area times the number of rotor lobes, times the stator pitch. A product of the input shaft rpm times the rotor lobes is called a "nutational" speed, which will be discussed in more detail below. Obviously, for a single-lobe rotor, nutational speed is equal to rotational pump shaft speed ($rpm_n = rpm \times 1$).

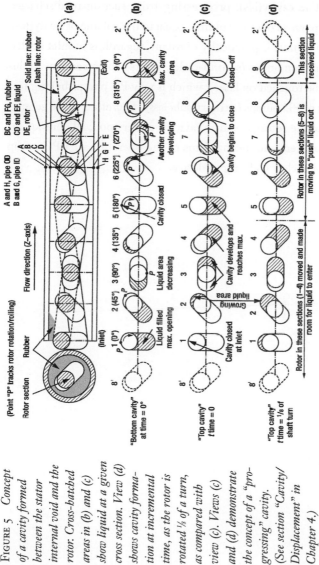

FIGURE 5 *Concept of a cavity formed between the stator internal void and the rotor. Cross-hatched areas in (b) and (c) show liquid at a given cross section. View (d) shows cavity formation at incremental time, as the rotor is rotated ⅛ of a turn, as compared with view (c). Views (c) and (d) demonstrate the concept of a "progressing" cavity. (See section "Cavity/ Displacement" in Chapter 4.)*

The cavity(ies), progressing from suction to discharge, pump out a constant volume, equal to the volume of cavity(ies). This volume is practically constant, regardless of inlet or discharge conditions (second-order factors and corrections to this will be touched on later), which is why PC pumps belong to a class of PD pumps (a rotary subclass). Therefore, if the geometry of the rotor-to-stator cavity is understood, all consequent parameters and mechanisms of the operation of a PC pump become clear and easy to derive.

Chapter 4
GEOMETRY

At first glance, the geometry of a PC pump hydraulic section (we will also apply this concept to DHPs and DHMs) could be discouragingly complex, which is probably why it is relatively poorly understood, even among users and within the engineering community. Not much information has been published that would explain the intricacies of the PC geometry in a clear, straightforward, and simple fashion. Consider, again, the assembled rotor/stator pair, as was shown in Fig. 5. Depending on the lobe ratio, the cross sections would look as shown in Fig. 6.

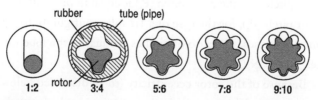

FIGURE 6
Cross-sectional view (perpendicular to pump axis) of various lobe ratios.

Because the number of rotor and stator lobes differ by one, a fluid-filled cavity is formed between a rotor and a stator. This is a key to its operating principle.

Cavity/Displacement

Consider the shape of the cavity for the simplest case, a single-lobe rotor (1:2 lobe ratio, i.e., two stator lobes). Several cavities can actually be seen in Fig. 5, as if frozen in time during the rotor rotation. For simplicity, a single stage is shown, but pumps usually have several stages, depending on the overall pressure differential between the suction and discharge. In Fig. 5, Sections 8′ and 2′ could be thought of as belonging to the adjacent stages, if there were any. The first (entrance) section shows the rotor positioned at the top extremity of the stator profile. As we move along the z-axis, the stator sections "turn" circumferentially until a complete 360-degree turnaround is reached at the other end, at the axial distance equal to stator pitch (P_s). To create an enclosed cavity, the rotor section must also "twist" around along the z-axis, but twice. The rotor completes its first 360-degree turnaround at the middle of the stator, and the second 360 degrees during the second half of the stator pitch (i.e., the rotor pitch (P_r) is equal to one-half of the stator pitch (P_s), for the case of a 1:2 lobe configuration). For multilobe cases, the ratio of pitches is equal to the lobe ratio (i.e., $P_s/P_r = N_s/N_r$).

Because of the rotor eccentricity (to be discussed below), the rotation of the rotor is similar to a car wheel when being driven along a road—the motion is actually an instantaneous rotation around the point of contact between the wheel and the road (and not around the center of the wheel). There is rolling with no sliding, as shown in Fig. 7. Point A generates a trajectory, called a *cycloid*.

If we examine positions of the stator sections in Fig. 5 at time $t = 0$, we can see several cavities. One of the cavities is starting from zero, opening at the top of the entry section (Section 1), and gradually opening as the section twists around—the small top cavity area at Section 2 (the 45-degree position). It grows slightly larger at Section 3, reaches the maximum at Section 5 (180 degrees, bottom of the stator), and then twists over on the other side. At the same time it is beginning to diminish, and eventually comes back to zero (the opening at Section 9), which is the end of the stator pitch (i.e., the end of one stage).

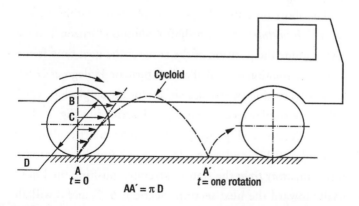

FIGURE 7
Rolling of a car wheel. Point B shows instantaneous rotation around point A. Distance AA′ is equal to the wheel's circumference.

This cavity ["top," Fig. 5(c)] is at maximum volume; it has zero opening at the inlet and exit sections and maximum opening in the middle of the stator pitch. At the same time, another cavity [Fig. 5(b)] occupies one-half of the stator pitch length, and has maximum opening at the inlet cross section, gradually diminishing toward the middle. From the middle, yet another cavity begins again and reaches maximum opening at the exit. Therefore, as the first cavity grows as we move along the axis, the second one becomes smaller and reaches zero volume just when the first one reaches its maximum, at Section 5. A third cavity starts off just where the second expires, and grows in a fashion similar to the first cavity, except that it starts at Section 5 (instead of Section 1, where the first cavity begins its growth). By the time the sections reach the end of the stator stage length (Section 9), the third cavity reaches its maximum size at the discharge cross section.

The above describes the position of a rotor within the stator *at a given time* (i.e., with shaft revolution "frozen"). If we now examine the position of the rotor at the next time frame during the rotation (e.g., ⅛ of a shaft turn, or 45 degrees), each rotor cross section (which are circles for the single-lobe case) will rotate similar to car wheels, as described in Fig. 7. The rotor in Section 2 [Fig. 5(d)] will move out a little, allowing the fluid to fill it, and so would Sections 3 and 4. Section 5 moves in, *away from* the stator extremity, pushing the liquid axially toward the next section. Sections 6, 7, and 8 will all perform similar movements, each receiving the same portion of fluid from the earlier sections, and at the same time displacing their own fluid toward the discharge. The cavity in Section

9 opens up to receive fluid from Section 8, while displacing its own fluid into the discharge. As shaft rotation continues, when the rotor reaches the lower stator extremity at Section 1, the cavity will travel one-half of its cycle (one-half shaft turn), and the first cavity will now be starting at the bottom in Section 1, reaching its maximum at the top of Cross Section 5, and diminishing back to zero at Section 9.

This process is continuous, resulting in smooth and continuous displacement of fluid from suction to discharge, similar to an auger inside a pipe. The movement of fluid is close to axial, but not exactly. In an auger, flow direction is straight due to the straight pipe ID, but the stator lobes deviate (guide the fluid) from the axially straight path, and its shape resembles a helix, which leads to helical motion of the pumped fluid. However, the rotational component of the fluid is small, which results in a low shear rate; this is why PC pumps are often used to transport fruits, vegetables, and other delicate products suspended in water with little damage.

If we "unwrap" the cavity from three dimensions and try to construct an equivalent two-dimensional representation, it will look as shown in Fig. 8, illustrating how pressure changes inside the cavity during the rotation cycle.

PC pumps are known to have low pulsations due to relatively straight flow path and low turbulence. However, there exists a small cyclical variation in pressure, as can be seen from Fig. 9, which reflects the progression of cavities inside the pump in accordance with rotor rotation.

Theoretically, fluid pressure changes instantaneously from suction to discharge. The result is a steplike change pressure

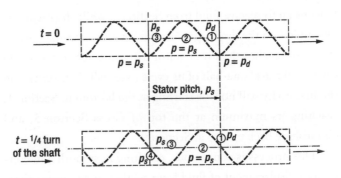

FIGURE 8

An "unwrapped" cavity. At top, cavity (1) is halfway with completing its travel through the stator and is at discharge pressure p_d. Cavity (2) is at its extreme position (same as "top" cavity shown in Fig. 5) and contains liquid still at suction pressure, just about ready to open up to discharge pressure. Cavity (3) is halfway inside the stator and is also at suction pressure. At bottom, shaft turned ¼ turn. Cavity (1) mostly completed its travel through the stator; cavity (2) opened up to discharge pressure (theoretically instantaneous pressure change); cavity (3) keeps growing into the stator; cavity (4) is just beginning to enter the stator.

fluctuation, alternating from suction to discharge values with a frequency of 1 × rpm for a 1:2 lobe case, or N_r × rpm for multilobes, in accordance with the concept of rotor nutations around a stator center. A concept of rotor nutations around a stator center is shown in Fig. 10.

Some pressure smoothing results due to interleakage as well as viscous friction. For multistage units, pressure fluctuations are based on per-stage values (i.e., the net pressure drop between suction and discharge divided by the number of stages). This reduces pressure pulsations even further. There is an interesting analogy between multilobe progressing cavity pumps (or mud-

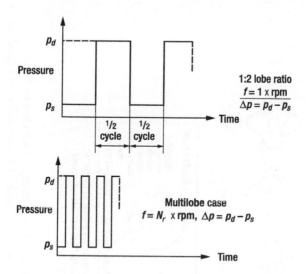

FIGURE 9
Pressure pulsations at a given point inside the pump, with the same magnitude but higher frequency as the lobe ratio is increased.

motors) and centrifugal pumps: for multilobe pumps, a mechanism of pressure pulsations is similar to "vane pass" frequencies characteristic of centrifugal pumps, caused by the impeller tips passing by the volute tongue or diffusor vane inlets.

Since pump unit displacement (q_o, flow per shaft revolution at zero load, indicated by the zero subscript) involves cavity volume (or several cavities for multilobe rotors), it is important to be able to calculate this volume. Even though the shape of the cavity is complex, there is a simple way to calculate its volume. The net fluid cross-sectional area (Fig. 5) is constant along the axis, and the cavity volume is equal to this fluid area times the cavity length (which is the stator pitch, P_s).

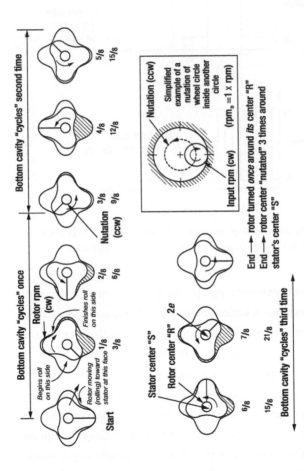

FIGURE 10   *Concept of rotor nutation for a 3:4 lobe ratio. During one rotation of the input shaft, the rotor "nutated" three times. Note that the direction of the nutation is opposite to the input shaft rotation. The insert shows a simplified case for a wheel (circle).*

As will be shown in more detail below, for multilobe rotors the result must be further multiplied by the number of rotor lobes to get the total displacement per one shaft revolution. This is because the cavity gets displaced N_r (number of rotor lobes) times during one revolution of the shaft (i.e., the rotor "nutations," as was shown in Fig. 10).

The mechanism described above can now easily be described with a few simple equations:

$$V_{cavity} = A_{fluid} \times P_s \text{ (in.}^3). \tag{1}$$

$$Q_o = V_{cavity} \times rpm_n = A_{fluid} \times P_s \times N_r \times rpm \text{ (in.}^3/min). \tag{2}$$

$$q_o = Q_o / rpm = A_{fluid} \times P_s \times N_r / 231 \text{ (gal/rev = gpr)}. \tag{3}$$

$$q_o = V_{cavity} \times N_r / 231 \text{ (gpr)}. \tag{4}$$

Unit flow displacement (q_o) depends only on the *geometry* of the hydraulics of the unit, but is independent of speed (rpm) and is a universal parameter for all positive displacement (PD) pumps, including rotary pumps such as gear, screw, lobe, etc. (Note: the above statement is true only for a theoretical or ideal pump, i.e., zero pressure differential or highly viscous pumpage, where the slip is zero or negligibly small.)

Profile Generation

Earlier, we discussed the simplest cycloid—a trajectory of a point on a car wheel as the wheel rolls along the road. However, the surface over which the wheel rolls (the circle) can have a finite curvature, as shown in Fig. 11. Only *hypocycloid* profiles (but not *epicycloid*) are used for generating PC pump profiles.

Unfortunately, there is no established nomenclature for the geometric parameters to designate profile characteristics

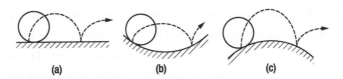

FIGURE 11
Cycloid (a); hypocycloid (b); epicycloids (c).

of PC units (or DHPs or DHMs). Different publications, countries, and manufacturers still use internal designations, which often leads to confusion and communication difficulties. Even such important parameters as major and minor diameters and pitch do not have associated symbols that are internationally accepted nomenclature. There is no standard, although several ISO committees are trying to bring some coherence to this industry. To consolidate these efforts, the nomenclature used in this book has been submitted to the ISO organization as a proposed International Standard, although it has not yet been published in completed form.

To begin profile generation, consider a *construction circle* of diameter *d*, with a *generating circle* with radius *e*, which rolls on the inside of the construction circle. There is no friction between the generating circle and the construction circle (i.e., pure rolling takes place). The resulting hypocycloid is shown in Fig. 12.

There is a certain restraining relationship between the construction and generating circle dimensions; otherwise, the generating circle would not end at the same starting position and no closed profile would be possible. The arc length (a) on

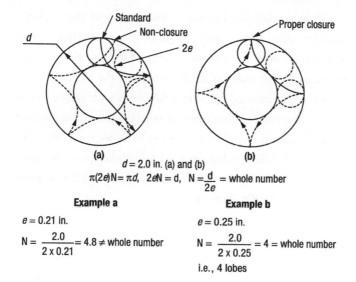

$d = 2.0$ in. (a) and (b)

$$\pi(2e)N = \pi d, \quad 2eN = d, \quad N = \frac{d}{2e} = \text{whole number}$$

Example a   **Example b**

$e = 0.21$ in.

$$N = \frac{2.0}{2 \times 0.21} = 4.8 \neq \text{whole number}$$

$e = 0.25$ in.

$$N = \frac{2.0}{2 \times 0.25} = 4 = \text{whole number}$$

i.e., 4 lobes

FIGURE 12

A restraining relationship must exist between the construction and generating circle dimensions; otherwise, the hypocycloid would not close.

the construction circle prescribed during one roll of the generating circle is equal to the circumference of the generating circle:

$$\pi \times (2e) \times n = \pi \times d.$$

There must be a whole number [$n = 2, 3, 4 \dots$ ($n = 1$ is a special case)] of such segments to fit inside the construction circle with no "leftover." In other words, the generating circle must roll inside the construction circle n times, such that

$$2ne = d. \tag{5}$$

This is a key fundamental relationship of the hypocycloidal profile of PC pumps. Figure 13 shows various hypocycloidal profiles for various values of *n*. The sharp cusps generated by this method are the centers of the *lobes*.

Lobes are created by adding *lobe circles* with radius *r* centered at the cusps. Therefore, *n* is equal to the number of lobes. When we talk about a 5:6 *lobe ratio*, we mean that the rotor has five lobes and the stator has six lobes (*n* = 1 is a special case). Figure 14 shows an example of such a 5:6 lobe ratio design (rotor profile shown).

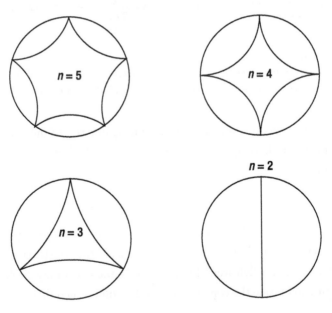

FIGURE 13
Various hypocycloids (before radius r has been added).

Next, the hypocycloidal profile is offset, normal to itself, by the distance *r*, which is same as the radius of the lobe circle shown in Fig. 15.

A smooth, continuous curve is thus created. The smoothness is important. When inspecting rotors, people often try to get a feel for the quality of the surface finish and shape by touching the nice, shiny chromed surface of a rotor. If a rotor was manufactured to correct specifications, no bumps or ridges should be detected by such a feel test. When bumps are

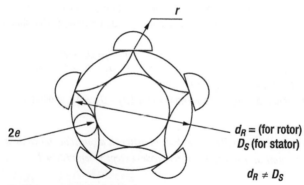

For rotor, $n = N_R$, $dr = 2eN_R$
For stator, $n = N_S$, $D_S = 2eN_S$
and also $N_S = N_R + 1$, therefore:

$$\frac{D_S}{d_R} = \frac{N_S}{N_R} = \frac{N_R + 1}{N_R} = \frac{N_S}{N_S - 1}$$

Note: Relationship between "*r*" and "*e*" can be arbitrary, but often taken as $r/2e = 1$, i.e. $r = 2e$

FIGURE 14
Lobe radius is added at each cusp of the hypocycloid, 5:6 lobe ratio design (rotor shown).

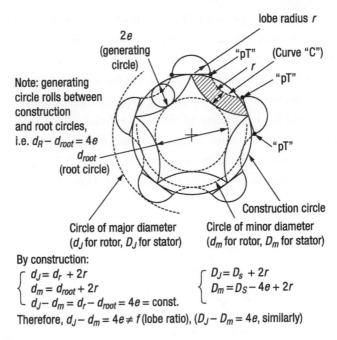

FIGURE 15

A completed PC profile. Note that, in general, a construction circle is not the same as a minor diameter circle, except when r/2e = 1.

detected, it is sometimes argued that the ridge on the rotor profile is an inherent property of the hypocycloid, at the point where the radius r translates into curve C, as shown in Fig. 15. Such claims are incorrect because point pT simultaneously belongs to a line that is normal to *both* the lobe circle and the offset hypocycloid (curve C), by the nature of its construction. This is a condition of curve continuity (smoothness). Such continuity is mathematically referred to as the "zero-order" continuity of a function. The higher orders are not continuous (first-order is curvature, etc.). Smoothness to the touch relates to the *function* itself, and not the curvature, so that the joining segments should indeed feel smooth along the entire surface if they were manufactured correctly.

From Fig. 15, the following relationships now become obvious, as shown below. Stator (number of lobes = N_s):

$$D_j = D_s + 2r. \tag{6}$$
$$D_m = (D_s - 4e) + 2r = D_j - 4e. \tag{7}$$
$$D_j - D_m = 4e, \text{ or } e = (D_j - D_m) / 4. \tag{8}$$

An important observation from Equation (8) is that the eccentricity e is equal to one-quarter of the difference between the major and minor diameters, *regardless of the number of lobes*. In practical terms, this makes it easy to calculate e and r based on measured values of just two parameters—the minor and major diameters. Substituting Equation (5) into Equations (6) and (7):

$$D_j = D_s + 2r = 2N_se + 2r. \tag{9}$$
$$D_m = (D_s + 2r) - 4e = D_j - 4e = 2N_se + 2r - 4e. \tag{10}$$
$$r = (D_j - 2N_se) / 2 = (D_m - 2N_se + 4e) / 2. \tag{11}$$

Rotor (number of lobes = N_r):

$$d_j = d_r + 2r = 2N_re + 2r. \tag{12}$$
$$d_m = (d_r + 2r) - 4e = d_j - 4e = 2N_re + 2r - 4e. \tag{13}$$
$$d_j - d_m \ (= D_j - D_m \ !) = 4e, \text{ or}$$
$$e = (D_j - D_m) / 4 = (d_j - d_m) / 4. \tag{14}$$
$$r = (d_j - 2N_re) / 2 = (d_m - 2N_re + 4e) / 2. \tag{15}$$

$$D_s / d_r = 2N_s e / 2N_r e = N_s / N_r = (N_r + 1) / N_r$$
$$= N_s / (N_s - 1). \tag{16}$$

Observe that eccentricity e and lobe radius r are the same for either the rotor or the stator, while the other parameters are part-specific. The core dimensions are X_m and X_j, with pitch equal to the stator pitch, P_s.

Next, let's calculate the flow area, A_{fluid}. First, a simplified (yet rather accurate) approach is to assume the areas between individual lobes to be approximately equal to the areas of the cavities between the lobes (i.e., the lobes are thought of as "flipped" cavities, as shown in Fig. 16). This approach is frequently utilized for other types of rotary pumps, such as gear

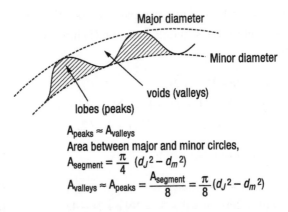

$$A_{peaks} \approx A_{valleys}$$

Area between major and minor circles,

$$A_{segment} = \frac{\pi}{4} \ (d_J{}^2 - d_m{}^2)$$

$$A_{valleys} \approx A_{peaks} = \frac{A_{segment}}{8} = \frac{\pi}{8}(d_J{}^2 - d_m{}^2)$$

Figure 16

Area approximation assumption, to calculate areas of rotor "peaks" and "valleys" (a similar assumption is often made to calculate the area between the teeth of rotary gear pumps).

types, where a tooth area is assumed to be approximately equal to the cavity between the teeth.

With such an assumption, the *shape* of the profile is irrelevant, since the net open area between the major and minor diameters is simply equal to one-half of the total area between the major and minor circles:

$$A_{valleys} = \pi / 8 \times (d_j^2 - d_m^2) \tag{17}$$

as shown in Fig. 16.

The fluid area is the difference between the stator cross-sectional opening minus the metal section of the rotor, per the calculation shown in Fig. 17:

$$A_{fluid} = \pi / 8 \times (D_j^2 + D_m^2 - d_j^2 - d_m^2). \tag{18}$$

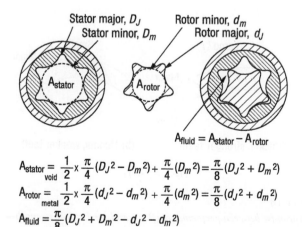

$$A_{\substack{stator \\ void}} = \frac{1}{2} \times \frac{\pi}{4}(D_J{}^2 - D_m{}^2) + \frac{\pi}{4}(D_m{}^2) = \frac{\pi}{8}(D_J{}^2 + D_m{}^2)$$

$$A_{\substack{rotor \\ metal}} = \frac{1}{2} \times \frac{\pi}{4}(d_J{}^2 - d_m{}^2) + \frac{\pi}{4}(d_m{}^2) = \frac{\pi}{8}(d_J{}^2 + d_m{}^2)$$

$$A_{fluid} = \frac{\pi}{8}(D_J{}^2 + D_m{}^2 - d_J{}^2 - d_m{}^2)$$

FIGURE 17
Fluid cross-sectional area illustration.

Substituting e and r in place of the major and minor diameters, we obtain

$$A_{fluid} = \pi/8 \times [(2N_s e + 2r)^2 + (2N_s e + 2r - 4e)^2 \\ - (2N_r e + 2r)^2 - (2N_r e + 2r - 4e)^2] = f(e, r). \quad (19)$$

In other words, knowing fundamental parameters e and r is sufficient to calculate net flow area. It also means that the same performance can be achieved with an infinite number of $r/2e$ ratios, for the same net *fluid area*, *pitch*, and *lobe ratio*. Historically, designers used this ratio as approximately $r/2e = 1$, although there are many designs with exceptions to that. $r/2e = 1$ is believed to be an optimum balance between the best performance (maximized flow area) and mechanical strength of a tooth shape, as shown in Fig. 18.

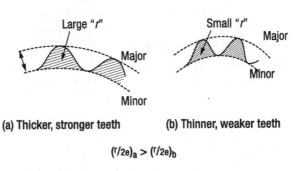

(a) Thicker, stronger teeth   (b) Thinner, weaker teeth

$$(r/2e)_a > (r/2e)_b$$

FIGURE 18
Compromise between performance (flow) and mechanical strength—thicker (a) versus thinner (b) profile teeth.

Single-Lobe (Special Case, $N_r = 1$, $N_s = 2$)

Most PC pumps have a lobe ratio of 1:2 (i.e., a single lobe rotor inside a two-lobed stator); there are some 2:3 ratio pump designs, but those are rare. Mudmotors, on the other hand, are usually multilobe designs with lobe ratios as high as 9:10. Since the stator of the 1:2 ratio design has two lobes, all previously derived formulas for multilobe cases apply to it. However, a rotor cross section is different. It is a simple circle whose diameter is a *minor* diameter, and is also equal to the stator minor diameter (i.e., $d_m = D_m$). A *major* diameter does not exist in the same sense as for multilobe configurations, where the major and minor diameters are in the same cross-sectional plane. Instead, a maximum diametrical dimension, encompassing the rotor, is used as the major diameter, although it is not in the same cross section. If projected into a cross section, it will appear as shown in Fig. 19.

All geometric relationships derived for the stator ($N_s = 2$) remain applicable here, but the rotor is different:

$$d_m = 2r. \tag{20}$$
$$d_j = 2r + 2e. \tag{21}$$
$$d_j - d_m = 2e \text{ (i.e., } \neq 4e \text{, as it was when } n > 1). \tag{22}$$

If a traditional value of $r/2e = 1$ is used, then

$$D_m = d_m = 4e. \tag{23}$$
$$D_j = 8e. \tag{24}$$
$$d_j = 6e. \tag{25}$$

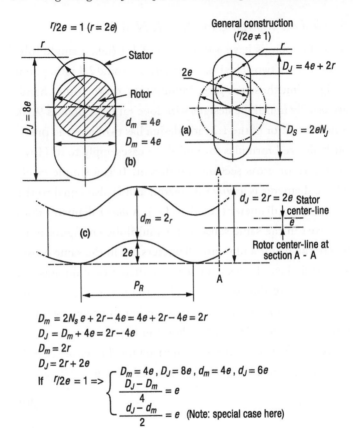

$D_m = 2N_s\,e + 2r - 4e = 4e + 2r - 4e = 2r$
$D_J = D_m + 4e = 2r - 4e$
$D_m = 2r$
$D_J = 2r + 2e$
If $\quad r/2e = 1 \Rightarrow$ $\begin{cases} D_m = 4e,\ D_J = 8e,\ d_m = 4e,\ d_J = 6e \\[4pt] \dfrac{D_J - D_m}{4} = e \\[6pt] \dfrac{d_J - d_m}{2} = e \quad \text{(Note: special case here)} \end{cases}$

FIGURE 19
Single-lobe case ($N_r = 1$, $N_s = 2$).

Rotor-to-Stator Fit (Clearance versus Interference)

Thus far, we have assumed a theoretical (zero) fit between the rotor and stator. For a real pump or mudmotor, this fit can be nonzero. To achieve less slip (higher *volumetric* efficiency, i.e., more pump flow), the rotor is made slightly larger diametrically, or the stator is made smaller, to create interference (i.e., restriction to slip). Typical examples of applications with tighter fit are for lower viscosity fluids that offer less viscous drag, and therefore more slip (fluid leakage) from high pressure back to low. However, if the fit is too tight, the friction between the metallic rotor and elastomer-lined stator would be excessive, requiring high operating and starting torque with reduced overall efficiency. The issue of high starting torque is very important, and is occasionally known to be a problem for PC pumps and mudmotors if fit is sized too tight. Premature wear, both gradual or catastrophic (called rubber "chunking"), is a sign of tight fit. There must be a compromise between flow and torque performance, and pump reliability and elastomer life.

Sometimes an artificially large clearance is intentionally designed in for higher temperature applications, to accommodate thermal growth at hot operating conditions where fit tightens. For such designs, the fit could be undersized by as much as 0.020–0.080 in., depending on the application conditions, elastomer properties, etc. Obviously, factory tests of such undersized designs do not have much meaning in such instances, because of the excessive "cold clearance" slip (i.e., significant loss of pumped flow). Factory tests for PC pumps at high temperatures are difficult and rarely conducted.

Radial fit varies between the major and minor values, varying along the sealing line. A decision as to which dimensions (rotor or stator) to alter to achieve desired fit is affected by various factors. For example, if rotor dimensions are kept theoretical, then the fit can be varied by using different cores for stator elastomer injection. This approach standardizes on rotors, and pump manufacturers must keep many cores on hand to produce "undersized" stators. Conversely, if a single core is used (standardizing on stator dimensions), then rotors with different fits need to be made for applications that require different fits. Each strategy has its merits and weak points. The approach of standardizing on stators allows for less trial and error development work with elastomers. Elastomers are always more difficult to fine tune, due to the high rate of thermal expansion, intricacies of the rubber vulcanization, and other process variables. Once the elastomer design work is completed, a manufacturer usually wants to "stick with what works" and may prefer to accommodate future fit changes by different rotor sizing. The production process of rotors is easier and more predictable. The user, on the other hand, might want to standardize on the lower-wear part (the rotor) to be able to try different stators for the same rotor for his or her applications.

In this book, for illustration purposes, we will assume the rotor dimensions do not change (remaining theoretical) and will accommodate fit requirements via stator size modifications. We will therefore add index (0, zero) for theoretical dimensions (i.e., d_{mo}, d_{jo}, D_{mo}, D_{jo}). Stator actual dimensions are therefore

$$D_j = D_{jo} + 2c_j. \tag{26}$$
$$D_m = D_{mo} + 2c_m \tag{27}$$
(but $d_j = d_{jo}$ and $d_m = d_{mo}$).

That is,

$$2c_j = D_j - D_{jo}. \tag{28}$$
$$2c_m = D_m - D_{mo}. \tag{29}$$

The above relationships establish a sign convention: a positive fit indicates clearance and a negative fit indicates interference. For example, +0.020 in. implies a 20-mil clearance (an enlarged stator made via a large core), and −0.030 in. implies a 30-mil interference (i.e., a smaller stator diameter made with smaller core).

The procedure to calculate the fit based on measured rotor and stator dimensions thus becomes simple. First, assume rotor theoretical dimensions equal to measured dimensions. Then, calculate theoretical dimensions for the stator. The difference between stator measured and theoretical dimensions is fit, positive for clearance and negative for interference.

Multilobe Case: Fit Calculations

$$D_{mo} = 2N_s e + 2r - 4e \text{ [per Eq. (10)]}.$$
$$2r = d_{jo} - 2N_r e \text{ [per Eq. (12)]}.$$

Then,

$$D_{mo} = 2N_s e + (d_{jo} - 2N_r e) - 4e = 2e \times (N_s - N_r) + d_{jo} - 4e$$
$$= d_{jo} - 2e \text{ (since } N_s - N_r = 1).$$

Substituting $e = (d_{jo} - d_{mo}) / 4$ into the above, we get

$$D_{mo} = d_{jo} - 2 \times (d_{jo} - d_{mo}) / 4 = (d_{mo} + d_{jo}) / 2.$$

That is,

$$D_{mo} = (d_{mo} + d_{jo}) / 2. \tag{30}$$

Next, for the major diameter:

$$D_{jo} = 2N_s e + 2r = 2N_s e + (d_{jo} - 2N_r e) = 2e \times (N_s - N_r) + d_{jo}$$
$$= 2e + d_{jo} \text{ (again, since } N_s - N_r = 1).$$

Substituting for e, we get

$$D_{jo} = (3d_{jo} - d_{mo}) / 2. \tag{31}$$
$$2c_m = D_m - D_{mo} = D_m - (d_{mo} + d_{jo}) / 2. \tag{32}$$
$$2c_j = D_j - D_{jo} = D_j - (3d_{jo} - d_{mo}) / 2. \tag{33}$$

As a reminder, Equations (32) and (33) are for the multilobe case (i.e., n > 1).

Single-Lobe (Special Case, $N_r = 1$, $N_s = 2$): Fit Calculations

The only difference in these calculations is that the eccentricity e is calculated differently than for a multilobe case:

$$e = (d_{jo} - d_{mo}) / 2 \ (\neq (d_{jo} - d_{mo}) / 4, \text{ as it was}$$
$$\text{in the multilobe case!}).$$
$$D_{mo} = 2N_s e + 2r - 4e = 2r = d_{mo} \text{ (since } N_s = 2 \text{ here).} \tag{34}$$
$$D_{jo} = 2N_s e + 2r = 4e + 2r = 4e + d_{mo}$$
$$= 4 \times (d_{jo} - d_{mo}) / 2 + d_{mo} = 2d_{jo} - d_{mo}.$$
$$D_{jo} = 2d_{jo} - d_{mo}. \tag{35}$$
$$2c_m = D_m - D_{mo} = D_m - d_{mo}. \tag{36}$$
$$2c_j = D_j - D_{jo} = D_j - 2d_{jo} + d_{mo}. \tag{37}$$

Equations (36) and (37) are for a single-lobe case, $N_r:N_s = 1:2$.

Core Sizing for Stator Production

This is very similar to any other injection process. For PC pumps and mudmotors, the injected material can be rubber, a rubber derivative, or other types of the elastomers (e.g., Viton®). The raw material comes in strips that are fed into a heated auger in the injection machine, and is moved by the piston under high pressure to fill the voids between the stator tube and a core that is positioned inside the tube and centered by the auxiliary equipment. The core is an exact duplicate of the stator "inverse image" stator (i.e., a stator "void" that has dimensions corrected for shrinkage). When heated to the rubber injection process temperature of approximately 350°F, the rubber undergoes vulcanization with consequent cooling in the autoclave. During this cooling process, the elastomer shrinks and separates away from the core, as shown in Fig. 20.

The original elastomer thickness is on the major and minor dimensions t_{mo} and t_{jo}. It is equal to the radial space between the tube ID and the core diameters X_j and X_m, which is occupied by the hot elastomer during injection. After cooling, the final rubber thickness is t_m and t_j.

The absolute shrinkages of minor and major equal diameters are

$$\Delta t_m = t_{mo} - t_m = (D_m - X_m) / 2. \tag{38}$$
$$\Delta t_j = t_{jo} - t_j = (D_j - X_j) / 2. \tag{39}$$

The relative shrinkage (shrink rate) is

$$s_m = \Delta t_m / t_{mo}. \tag{40}$$
$$s_j = \Delta t_j / t_{jo}, \text{ and these vary depending on the elastomer.} \tag{41}$$

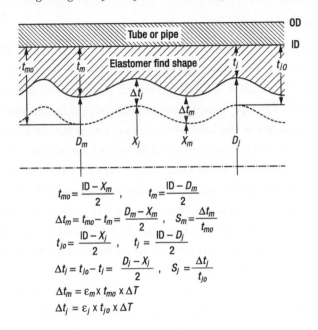

$$t_{mo} = \frac{ID - X_m}{2} \; , \qquad t_m = \frac{ID - D_m}{2}$$

$$\Delta t_m = t_{mo} - t_m = \frac{D_m - X_m}{2} \; , \quad S_m = \frac{\Delta t_m}{t_{mo}}$$

$$t_{jo} = \frac{ID - X_j}{2} \; , \qquad t_j = \frac{ID - D_j}{2}$$

$$\Delta t_j = t_{jo} - t_j = \frac{D_j - X_j}{2} \; , \quad S_j = \frac{\Delta t_j}{t_{jo}}$$

$$\Delta t_m = \varepsilon_m \times t_{mo} \times \Delta T$$

$$\Delta t_j = \varepsilon_j \times t_{jo} \times \Delta T$$

FIGURE 20
Dimensional changes during stator production.

The thermal expansion rate for the elastomers is

$$\varepsilon_m = (\Delta t_m / (t_{mo} \times \Delta T), \text{ in. } / (\text{in.} \times {}^\circ F). \tag{42}$$

$$\varepsilon_j = (\Delta t_j/(t_{jo} \times \Delta T), \text{ in. } / (\text{in.} \times {}^\circ F). \tag{43}$$

For the elastomers, a thermal rate of expansion is an order of magnitude greater than that for metal, which is why the thermal changes in metal are neglected in these calculations.

Thermal expansion of elastomers used for PC pumps and mudmotors is complex. The elastomer is free to expand radially, but is restricted in the axial direction due to its attachment (bond) to the tube ID. The net result is that the radial expansion rate is somewhere between the linear and volumetric.

The thermal coefficient can be approximated as: $\varepsilon_V \approx 3 \times \varepsilon_m$. The linear thermal expansion coefficient rate for rubber is approximately $\varepsilon_m \approx 70 \times 10^{-6}$ in. / (in. $\times$ °F) (i.e., ten times that of metal). The volumetric rate would then be about three times greater, $\varepsilon_V = 200 \times 10^{-6}$ in.^{3} / (in.$^3 \times$ °F). Conservatively, a suggested value for the linear coefficient is 60×10^{-6} in. / (in. $\times$ °F) for regular BUNA rubber, and approximately 200×10^{-6} in. / (in. $\times$ °F) for HSN (high saturated nitrile). However, values as high as 470×10^{-6} in. / (in. $\times$ °F) are reasonable for some elastomers.

A wavy shape of the elastomer profile results in a thinner layer of rubber on the major dimension and a thicker layer on the minor dimension. This would imply that the elastomer's absolute growth is proportionally greater on the minor dimension than on the major (for the same *relative* rate of growth). However, interactions within the cross section (as well as in the lateral direction) modify this process. Thinner material at the major dimension gets pulled to grow more, and thicker sections on the minor dimension are restrained and grow less due to internal stresses. This is why the values of thermal expansion coefficients are not the same on the minor and major dimensions, but are established empirically, accounting for design specifics. Finite element analysis (FEA) is a good engineering tool for evaluating the thermodynamic mechanisms of elastomers and to size parts appropriately for the actual operating conditions. The most reliable results, however, can be best obtained through actual test data. As one gains more experience in the design and applications of PC pumps and mudmotors, predicting and sizing of these units for proper fits at operating temperatures will become more accurate and reliable.

Performance: Operating Characteristics

As was discussed earlier, *unit flow* (q_o = Q/rpm) is an important characteristic parameter describing the flow per revolution of the drive shaft (i.e., gallons per revolution, gpr). This characteristic is fundamental for any rotary or other positive displacement (PD) pump. Since q_o is related to the geometry of the cavity, and a simplified method to estimate its volume was shown earlier [see Equations (3) and (4)], a pump flow at zero pressure is

$$Q_o = q_o P_s N_r / 231. \tag{44}$$

If the differential pressure across the pump is not zero, corrections for slip must be made. However, for high and moderate viscosity fluids this correction is small, and Equation (44) can still be used as an approximation.

Accounting for slip, as illustrated in Fig. 21, the pump flow and volumetric efficiency are

$$Q = Q_o - Q_{slip}, \text{ and } \eta_{vol, pump} = Q / Q_o. \tag{45}$$

For a mudmotor, this is

$$Q = Q_o + Q_{slip}, \text{ and } \eta_{vol, MM} = Q_o / Q. \tag{46}$$

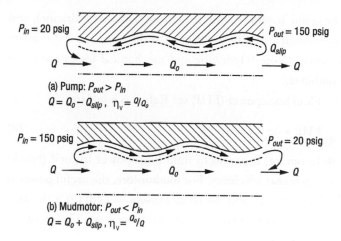

FIGURE 21
Slip (leakage) and volumetric efficiency for PC pumps and mudmotors.

As is similar with all positive displacement machines (flow proportional to speed), the torque is equal to

$$T = (Q/rpm) \times (\Delta p \times 231 / (24\pi) \times \eta$$
$$= q \times (\Delta p \times 231) / 24\pi) \times \eta. \qquad (47)$$

In the United States, the units used are: flow in *gpm*; rotational speed in *rpm*; pressure in *psi*; and torque in *ft* $\times$ *lbs*. Overall efficiency (η) is less often calculated with these types of pumps, as compared with other types (e.g., centrifugal). Instead, volumetric efficiency has been a traditional parameter for comparing these machines. Nevertheless, overall efficiency is an important consideration when comparing PC pumps to other pump types. It takes into account all losses, including those due to slip (volumetric), fluid friction (hydraulic), and mechanical losses such as friction at the rotor and stator interference. (Note: Hydraulic and mechanical losses are often combined.)

Fluid horsepower (FHP, see Ref. 3) is equal to

$$FHP = \Delta p \times Q / 1,714. \tag{48}$$

In pumps, the ratio of fluid horsepower to total (break) horsepower is efficiency. For mudmotors, the useful power is hydraulic power, and the break horsepower (BHP) is the supplied power (i.e., it is greater than fluid power). Therefore,

$BHP = FHP / \eta$, for pumps,
$BHP = FHP \times \eta$, for mudmotors.

Mechanical torque is related to horsepower via

$BHP = T \times rpm / 5,252.$

From the above equations, we see that torque depends on differential pressure across the machine (pump or mudmotor), and is independent of speed or flow, since $q = Q/rpm$ is constant (neglecting the slip adjustments). Examples of pump and downhole mudmotor performance curves are shown in Fig. 22.

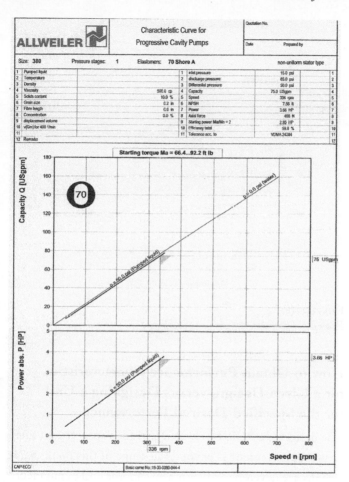

FIGURE 22(A)
Example of performance curves: PC pump.

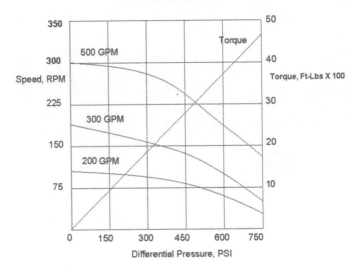

FIGURE 22(B)
Example of performance curves: mudmotor.

How to Obtain Performance Characteristics for a Given Design, versus Designing a Unit for the Specified Desired Performance

The design of a PCP, DHP, or DHM for a specific customer or user requirement is beyond the scope of this book. Such designs are developed by pump and motor design engineers with the use of computer programs, empirical procedures, factory tests, and field reliability feedback (see Ref. 4).

However, the evaluation of performance based on known geometry is simpler, requiring only a few basic dimensions of a pump or a mudmotor. If the major and minor dimensions of the rotor and stator are measured, and the stator pitch is known, a complete set of operating characteristics can then be predicted with reasonable accuracy, as shown in the following examples.

Examples

Example 1: Performance Estimate from the Given Measured Geometry

Consider a mudmotor design for an oil-well drilling application. The rotor is made from chrome-plated steel and the stator tube is lined with BUNA-N 70 durometer rubber. The tube OD size is 6.75 in. for this five-stage unit with a lobe ratio 5:6. It is intended to operate at 212°F fluid temperature. The unit has the following dimensions:

Stator:
$D_m = 3.762$ in.
$D_j = 5.055$ in.
$P_s = 26.666$ in.

Rotor:
$d_m = d_{mo} = 3.087$ in. (assumed to be equal to theoretical)
$d_j = d_{jo} = 4.388$ in.
$P_r = 22.17$ in.
$e = (d_{jo} - d_{mo}) / 4 = (4.388 - 3.087) / 4 = 0.325$ in.
$r = (d_{jo} - 2N_re) / 2 = (4.388 - (2 \times 5 \times 0.325)) / 2 = 0.568$ in.
$r/2e = 0.568 / (2 \times 0.325) = 0.87$ in. (somewhat less than
the "traditional" $r/2e = 1$)

Core (core dimensions would not be known to a user in the field, but are given here for illustration):

$X_m = 3.684$ in.
$X_j = 5.026$ in.
$P_{core} = P_s = 26.666$ in.

Applying previously derived equations, we will first calculate rotor/stator fit in cold (i.e., ambient) conditions:

$D_{mo} = (d_{mo} + d_{jo}) / 2 = (3.087 + 4.388) / 2 = 3.738$ in.
[per Eq. (30)].
$D_{jo} = (3d_{jo} - d_{mo}) / 2 = (3 \times 4.388 - 3.087) / 2 = 5.039$ in.
[per Eq. (31)].
$2c_{m,cold} = D_m - D_{mo} = 3.762 = 3.738 = +0.025$ in.
$2c_{j,cold} = D_j - D_{jo} = 5.055 - 5.039 = +0.017$ in.

Note that there is a clearance (plus sign) between the rotor and the stator prior to their being exposed to the actual operating temperatures.

To predict performance at actual operating temperatures, we need to determine the linear coefficient of expansion. A value of 60^{-6} in. / (in. $\times$ °F) is typical for BUNA. However, since we happen to have information on core dimensions (which is usually not available in the field), we can better evaluate the elastomer's thermal behavior. Figure 23 illustrates a mechanism of thermal changes during the stator production at the factory. From this data, we will calculate the rubber shrinkage based on the difference in elastomer thickness at injection temperature (when it fills the void between tube ID and a core), and the elastomer thickness at ambient conditions.

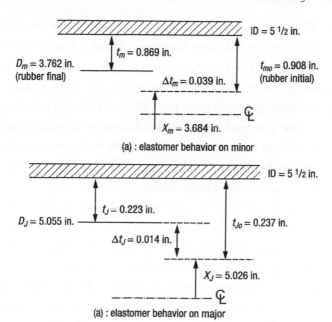

FIGURE 23
Elastomer shrinkage during the injection process (350°F), and during cooling to ambient (80°F). ΔT = 350 − 80 = 270°F.

The process temperature is around 350°F and ambient is assumed to be 80°F [refer to Equations (38) through (43)].

From Fig. 23, based on minor dimensions:

$$\varepsilon_m = (\Delta t_m) / (\Delta T \times t_{mo}) = 0.039 / [(350 - 80) \times 0.908]$$
$$= 159^{-6} \text{ in. } / (\text{in.} \times °F).$$

And based on major dimensions:

$$\varepsilon_j = (\Delta t_j) / (\Delta T \times t_{jo}) = 0.014 / [(350 - 80) \times 0.237]$$
$$= 219^{-6} \text{ in. } / (\text{in.} \times °F).$$

The empirical value of 60^{-6} in. / (in. $\times$ °F) is below the values for the minor and major dimensions. Which value, or values, should we use? The following considerations are recommended in such situations. For old, proven designs, the actual process-based shrinkage would seem to be a better choice. This is because these data would be based on a large number of produced units, offering enough statistical data to form a better basis for the thermal coefficient values, which will be different for the minor (ε_m) and major (ε_j) dimensions.

However, statistically valid samples for new designs have not yet been accumulated, and initial process errors and fluctuations may bias the results. In such cases, it is better to use the more general, established values for the types of elastomers used in the past for similar designs. We will therefore use 60^{-6} in. / (in. $\times$ °F) to evaluate the elastomer behavior in the field, at an actual operating temperature of 212°F.

$$\Delta t_m = 0.908 \times (212 - 80) \times 60 \times 60^{-6} = 0.007 \text{ in.}$$
$$\Delta t_j = 0.223 \times (212 - 80) \times 60 \times 60^{-6} = 0.002 \text{ in.}$$

The operating fit (at application temperature) would then be

$$2c_{m,hot} = +0.025 - 2 \times 0.007 = +0.011 \text{ in.}$$
$$2c_{j,hot} = +0.017 - 2 \times 0.002 = +0.013 \text{ in.}$$

The issue of what constitutes a proper operating fit is still debated in the industry, and depends on the pump size, the elastomer, the chemistry of the pumped fluid, the starting and operating torques, reliability and life considerations, and other factors. Usually, a lighter fit is a better starting approach, with the design being fine-tuned after actual field data becomes available. This is a safer and more conservative approach. Should a problem arise, it would likely be low torque, but the unit would still operate and not fail prematurely. On the other hand, if the fit is too tight initially, it may cause catastrophic chunking or starting problems.

In our example, the operating temperature of 212°F is only an estimate. For a mudmotor operating at depths of thousands of feet, it is difficult to know the exact temperature conditions before actual drilling starts. If the mud temperature proves to be higher than the estimate, a tighter operating fit would result; if the temperature is lower, the design can be adjusted during follow-up design upgrades to tighten the fit. Therefore, when the operating conditions have not yet been verified, a lighter fit is better for the initial tests.

For calculating performance estimates, the cavity area is [from Equation (18)]

$$A_{fluid} = \pi/8 \times (D_j^2 + D_m^2 - d_j^2 - d_m^2)$$
$$= \pi/8 \times (5.055^2 + 3.762^2 - 4.388^2 - 3.087^2)$$
$$= 4.29 \text{ in.}^2$$

The unit flow is then

$$q_o = A_{fluids}P_sN_r/231 = 4.29 \times 26.666 \times 5/231 = 2.40 \text{ gpr.}$$

Maximum shaft speed selection depends on many factors, and is a function of tube size, the number of lobes, and the operating conditions. It must be selected based on experience.

Generally, lower speeds are recommended (see Ref. 5) for mudmotors to increase MTBF (mean time between failures, i.e., longer life), although the production (drilling rates) would suffer somewhat. Higher speeds are used when the drilling speed is critical (e.g., if the job is estimated to be quick and the total production time is below the mudmotor's MTBF). In such cases, the elastomer life is sacrificed in favor of production time. The overall decision must be based on economics— initial cost of equipment and time (i.e., cost) if units fail for pullouts and changeovers, etc. For our example, 250 rpm was empirically chosen as a maximum value at full load. This speed will be higher at zero load, due to volumetric inefficiency (slip), as shown below.

Typical volumetric efficiencies of mudmotors are in the range of 80–90% (see Ref. 6). Higher values may indicate potential problems with wear, while lower values would mean insufficient fit. Using 80% as a ballpark estimate, we get

$$rpm_o = rpm / \eta = 250 / 0.8 = 312 \text{ rpm.}$$
$$Q_o = q_o \times rpm = 2.40 \times 312 = 740 \text{ gpm.}$$

Thus, 740 gpm must be delivered to the mudmotor by the charging equipment at the surface.

Note: For mudmotors, slip results in less flow through the cavity, causing slower rotational speed at full load. In contrast, for pumps the speed is constant and equal to the driver speed, and the pump flow decreases with load. In other words, in mudmotors, flow is the cause for rotation, while in pumps the rotation leads to flow.

Overall maximum pressure drop is based on the empirically established pressure differential per stage. This value is 125 psi for mudmotors and 75 psi for pumps. The higher allowable values for mudmotors are due to the fact that in drilling applications, the time to accomplish the job is typically much more critical than for surface-mounted PC pumps. It is expensive and puts greater demands on the downhole equipment. A typical oil-well drilling operation takes one to three months, while a typical PC pump installed at, for example, a paper mill may need to last for many years, and often has a spare standby unit.

The overall length and diameter of mudmotors is more critical than in equivalent surface pumps (Ref. 7), due to size constraints of the drilled hole. This is another reason to stretch the limits for more pressure, torque, and power in shorter lengths. In our example, a five-stage (five stator pitches) design would allow $5 \times 125 = 625$ psi differential pressure. If such calculations are performed in the same way at different speeds, the resulting performance characteristics can be plotted, similar to Fig. 22.

Finally, assuming a 75% overall efficiency, the torque is

$$T = q \times \Delta p \times 231 / (24\pi) \times \eta$$
$$= 2.40 \times (625 \times 231 \times 0.75) / (24 \times 3.14)$$
$$= 3,550 \text{ ft} \times \text{lbs.}$$

Example 2: What Happens if a Mismatched Rotor-Stator Pair Is Combined?

Let's say that in response to an urgent call from a customer, a manufacturer's production planning manager searches the inventory, trying to find a mudmotor power section available for a quick delivery. There are several extra stators available (same as in Example 1), but there are no more rotors. However, a rotor with the same pitch dimension but with somewhat different major and minor diameters is available:

d_{mo} = 3.133 in. (versus required 3.087 in.).
d_{jo} = 4.343 in. (versus required 4.388 in.).

Could this rotor be still used for the application conditions as described in Example 1? Calculating the cold fit:

$2c_{m,cold}$ = 3.762 − (4.343 + 3.133) / 2 = +0.024 in.
$2c_{j,cold}$ = 5.055 − (3 × 4.343 − 3.133) / 2 = +0.107 in.

For hot conditions (212°F temperature of the mud at the mudmotor position), the operating fit will be

$2c_{m,hot}$ = 0.024 − 2 × 0.007 = +0.010 in.
$2c_{j,hot}$ = 0.107 − 2 × 0.002 = +0.103 in.

(Note that the hole environment temperature may be different from the mud temperature at the power section; this point is subtle and often missed.)

There appears to be a significant difference between the major and minor fit, and the clearance (+0.103 in.) at the major dimension seems to be excessive. If anything, the fit at the major should be less than that on the minor (to produce somewhat equal relative fits, based on equal-stress theory, i.e., thicker elastomer sections at the minor grow more and should have proportionally tighter interference than at the major):

$c_m / c_j \sim t_{mo} / t_{jo}$, or $c_m / t_{mo} \sim c_j / t_{jo}$.

Of course, this theory would apply to cases with interference (negative fits), and not for clearances, which is the case here. Should temperatures in the drilled hole be higher than estimated, the fits would tighten up—the minor dimension would produce an interference fit, and the major dimension may still remain with clearance—with in-between points along the seal line. The end result is difficult to predict. The ultimate application conditions may require design modifications. For example, abrasion wear may become a problem as more information about the application becomes available, and the speed might need to be reduced (Refs. 8, 9). Perhaps the unit could be offered on a trial basis, but no guarantees could be extended because of unknown performance.

Multilobe versus Single-Lobe Geometry: More Performance?

Operators of drilling rigs know that multilobe mudmotor designs are stronger, producing more torque and more power compared to fewer-lobe designs. The greater the number of lobes, the stronger the mudmotor is believed to be. And this is true—an 8:9 lobe configuration will give more flow and torque than a 2:3 lobe configuration. Why?

It is difficult to find an actual installation example where one could strictly compare only the effect of lobe ratio changes. Other parameters are usually different, as well, such as pitch, interference, etc. This is because the higher number of lobes allows lower pitch for about the same flow and speed, resulting in savings on the overall unit length. To truly under-stand the reasons behind the effect of lobe ratio alone, we must freeze all other parameters, having the same input rpm, same pitches, same interference, same $r/2e$ ratio, and also using the same tube. Using the same tube ID implies the same stator major diameter (to maintain the same minimum required rub-ber thickness). This would require changes to the stator minor dimension, as well as to both rotor minor and major dimen-sions. These changes will affect the net flow area, which, together with the newly selected number of lobes, will produce different unit flow (q) and, therefore, different torque.

The mathematics of this are relatively easy, although sel-dom demonstrated in the literature, and follow from the equa-tions derived earlier for the geometry calculations and performance assessment. Consider the 5:6 lobe ratio design discussed in Example 1:

Stator (tube ID = 5.5 in., N_s = 6):
D_{jo} = 5.039 in. (theoretical, as calculated in Example 1)
D_{mo} = 3.738 in.
P_s = 26.666 in.
e = 0.325 in.
r = 0.568 in.
$r/2e$ = 0.87 (less than the traditional $r/2e$ = 1)

Rotor (N_r = 5):
d_{jo} = 4.388 in.
d_{mo} = 3.087 in.

Let's examine what happens if the lobe ratio is changed to 9:10 ($N_r = 9$, $N_s = 10$). Originally, in Example 1, this mudmotor unit flow was q = 2.40 gpr. Since the lobe number changed, the eccentricity would also change, in accordance with previously derived geometric relationships (Equation (9), rearranged):

$$e = D_{jo} / (2 \times (N_s + 2\ r/2e)) = 5.039 / (2 \times (10 + 2 \times 0.87))$$
$$= 0.214 \text{ in.}$$

As was mentioned earlier, for consistency, we kept the same ratio of $r/2e = 0.87$. Therefore,

$$r = (r/2e) \times 2e = 0.87 \times 2 \times 0.214 = 0.376 \text{ in.}$$

The changed eccentricity forces a change of the stator minor diameter. It will increase, at first helping to open up the flow area:

$$D_{mo} = D_{jo} - 4e = 5.039 - 4 \times 0.214 = 4.183 \text{ in.}$$

Next, the new rotor dimensions, to match a new stator, will become

$$d_{jo} = 2N_r e + 2r = 2 \times 9 \times 0.214 + 2 \times 0.376 = 4.604 \text{ in.}$$
$$d_{mo} = d_{jo} - 4e = 4.604 - 4 \times 0.214 = 3.748 \text{ in.}$$

The net fluid area is

$$A_{fluid} = \pi/8 \times (5.039^2 + 4.183^2 - 4.604^2 - 3.748^2) = 3.0 \text{ in.}^2$$

Finally, the unit flow is

$$q_o = A_{fluid} P_s N_r / 231 = 3.0 \times 26.666 \times 9 / 231 = 3.12 \text{ gpr.}$$

The net unit flow increased ($3.12 / 2.4 \sim 1.3$, i.e., 30%) but not directly with the increase of number of rotor lobes, for the following reason. The number of rotor lobes did cause a strong effect on the net fluid area ($9/5 = 1.8$), but the net cross-sectional area A_{fluid}, however, actually decreased ($3.0/4.15 = 0.72$). Since the unit flow is affected by both of these, the net increase in q_o is a combination of effects: the increased number of lobes has a stronger effect on flow (increases it), while the reduced net cross-sectional area tends to decrease the flow (i.e. retarding the net flow increase somewhat). In essence, a "nutational" effect of the rotor is evident: for the same rpm of the input shaft, the rotation (i.e., nutation) of the rotor around the stator's center is multiplied by rpm_n, which results in (rpm_n/rpm) more flow, but moderated by the area correction.

From the standpoint of the relative effect of different rotor and stator parameters, it might be of interest to observe that while the stator major diameter was fixed ($5.039/5.039 = 1$), the minor diameter increased by $4.183/3.738 = 1.12$ (12%); the rotor major diameter increased by $4.604/4.388 = 1.05$ (5%), and the rotor minor diameter increased by $3.748/3.087 = 1.21$ (21%). In other words, the growth of the rotor dimensions occurred faster than the stator minor dimensions, blocking the fluid area and negating the effect of the area increase via the stator minor dimensions, while the stator major dimensions remained passive.

In principle, if we continue to increase the lobe ratio to infinity, the contours of the rotor and stator cross sections

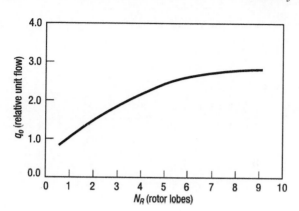

Relationships between unit flow (q_o), torque (T), versus number of lobes.

would begin to approach circles (Ref. 10). Any further changes in area would obviously be negligible, while, for example, another doubling of the rotor lobes would double the unit flow, q_o. If the results are plotted for unit flow (and, proportionally, torque) versus different lobe numbers, the relationship shown in Fig. 24 becomes evident.

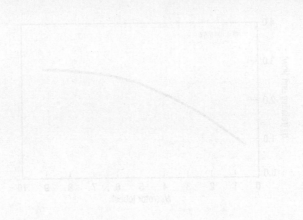

Chapter 5

DESIGN-RELATED CONSIDERATIONS

In Chapter 4, we examined progressing cavity machines from the standpoint of the fundamental principles of their construction and operation. We established geometric relationships between eccentricity, pitch, construction circle, minor and major diameters, etc., as well as the nomenclature of consistent definitions. These relationships were derived from those fundamental first principles to form a solid foundation for application guidelines of these machines.

Chapter 5 will compare these application guidelines and interlocking theoretical principles with empirical pump designer- and user-based experiences, which ultimately form the ever-improving database of knowledge and understanding of the optimum application of these units in the field.

Key Design Parameters

Of all the design parameters covered in Chapter 4, only a few critical ones are required from an applications standpoint. These geometrical variables are: *eccentricity*, the *minor* and *major* diameters for rotors and stators, and the rotor and stator *pitches*. Figure 25 shows the geometry of a typical stator of a PC pump. Refer to Fig. 15 and the accompanying text in Chapter 4 for the exact derivation of the relationships among the parameters of the rotor-stator geometry, and the difference between the theoretical (zero-fit) and actual diameters.

In Chapter 4, in our discussion of sizing stator core dimensions, we touched briefly on the stator elastomer thickness and the changes in thickness due to thermal expansion at operating conditions or during the manufacturing process. We did not discuss determining the actual *value* of the thickness of the

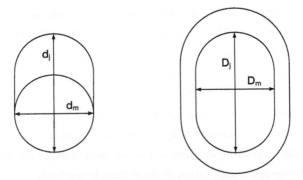

FIGURE 25
Geometry of a typical stator of a PC unit (courtesy Colfax Pump Group).

elastomer, but from the design and applications standpoint this is very important. The thickness of the elastomer should be sufficient to cushion and minimize the stresses caused by the rotor sliding against the stator during operation, especially for cases with significant interference. On the other hand, if the elastomer lining is too thick it may compress too much, resulting in poor performance due to leakage (slip) of flow from the discharge back to suction, and inability of the too "spongy" elastomer to resist this leakage. Therefore, in addition to the purely geometric parameters (eccentricity, minor and major diameters, and pitches), the elastomer thickness is another critical parameter that the designer must consider for the pump application's specific operating conditions.

Other key parameters are *flow area*, *cavity volume*, *average fluid velocity*, *maximum particle velocity*, and *maximum rubbing velocity*. Some of these were derived in Chapter 4, although some approximating assumptions were made. For example, flow area (see Fig. 16) was derived assuming that the areas of the peaks and valleys were equal, which allows a small margin of error. Due to the complex shapes of PC machines, the better way is to physically measure the net fluid area of the rotor/ stator cross section after the design is completed.

Consideration of particle velocity, on the other hand, is strictly empirical, based on years of history and experience that have produced solid guidelines for various operating conditions and construction materials. Therefore, fluid velocity, rubbing velocity, and particle velocity are selected to produce geometries resulting in sufficient PC machine life and good reliability. The three variables that most significantly affect these velocities are eccentricity, minor and major diameters, and pitch. Figures 26 through 28 illustrate the concept of the three velocities in a PC pump.

As can be seen from Fig. 27, the rubbing (surface) velocity is a sum (or a difference) of two components: rotational velocity, which is constant, and a traversing velocity, which is variable. It reaches maximum at the middle position (center), and

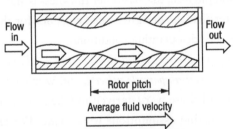

Flow in

Flow out

Rotor pitch

Average fluid velocity

The average speed at which the product travels though the pipe

$$A.F.V. = \frac{Flow\ rate}{Flow\ area} = \frac{Q \times 0.321}{A_{fluid}} = \frac{Q \times 0.321}{4ed_m} \quad (ft/sec.)$$

d_m

Flow area $= 4ed_m$

$4e$

- The average fluid velocity through the pump element is important because of it's influence on wear, too high a velocity causes erosive wear and too low a velocity allows solids to settle and abrasive wear occurs.

FIGURE 26
Average fluid velocity (AFV).

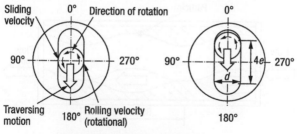

The surface velocity (at any position) =
rotational velocity ± traversing velocity
    (constant)   (varies)

The maximum rubbing velocity occurs at the "A" position

- The velocity with which the rotor rubs against the stator has a very significant effect on the wear rate of both components.
  The rubbing velocity is not constant as the rotor revolves, it has maximum and minimum values at different points around the stator slot.

Peripheral rubbing velocity is the linear distance around the circumference of the stator slot divided by the time the rotor takes to make one revolution

$$V_{surface} = V_{rotational} \pm V_{traversing}$$

FIGURE 27
Rubbing (surface) velocity.

reduces to zero at the extremes (the top and bottom positions). However, the average value of the traversing velocity is simply the distance that the rotor traverses back and forth during each revolution times the number of revolutions per minute (rpm). This distance traversed per each revolution is equal $4e \times 2$ (up and down). Rotational velocity is simply a product of the angular speed times the radius (one-half of the minor diameter). In U.S. units (using dimensions in inches with appropriate conversion coefficients), it is easy to show that

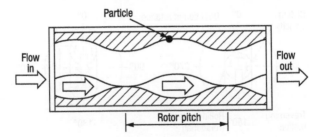

The maximum velocity of a particle travelling at the extremes
The stator major diameter

Maximum particle velocity =

$$\frac{\text{rpm} \times \sqrt{\pi^2(4e + d_{mo})2 + P_R^2}}{229} \text{ (ft /sec)}$$

• Maximum particle velocity is an important criteria, this influences the viscous liquid handling and cavitation characteristics.

FIGURE 28
Maximum particle velocity.

$V_{rot} = \text{rpm} \times d_m / 229 \text{ ft/sec.}$

$V_{tran} = 8e \times \text{rpm} / 12 \text{ ft/sec.}$

$V_{surf} = \text{rpm} \times d_m / 229 \pm 8e \times \text{rpm} / 12$

$\quad\quad = \text{rpm} \times (d_m / 229 \pm 8e / 12) \text{ ft/sec.}$

For maximum particle velocity, Fig. 28 tracks the particles that move at the farthest extreme away from the centerline (the highest rotational motion). It also covers the distance that the rotor carries it through (i.e., the rotor pitch).

Important Ratios

As we explained in Chapter 4, construction of the geometric profile of a PC machine starts with selecting a construction circle d (d_r for the rotor and d_s for the stator). As was shown in Fig. 15, the eccentricity (e) and lobe radius (r) are chosen independently, and the proportions between r and e create geometries with either a shallower profile or with more pronounced peaks, as was shown in Fig. 16. The ratio $r/2e$, therefore, is a measure of how peaked the profile is. Most traditional designs use $r/2e = 1$, but not always.

Once the design is finished and actual units are manufactured, the construction diameter is no longer accessible for direct measurement; only the major and minor diameters and the pitch can now be measured. The language of the designer ($r/2e$) is replaced by the language of the application engineer (d_m/e). Examining Fig. 15, it can be easily shown that

$$d_m/e = (d_{root} + 2r)/e = (d - 4e + 2r) \, / \, e = d/e - 4 + 4(r/2e).$$

Or, substituting for $d = 2eN_r$, we get

$$d_m/e = 2eN_r/e - 4 + 4(r/2e) = 2N_r - 4 + 4(r/2e).$$

For a given lobe ratio design, there is a direct relationship between the ratios $r/2e$ and d_m/e. For a given nominal pump size (i.e., having a given d_m), choosing the ratio d_m/e sets the eccentricity, e.

As was shown in Fig. 19, for a special case of a single-lobe design (most PC pumps are single-lobed, $N_r = 1$):

$$d_m = 2r.$$

When the lobe radius r is set equal to the diameter of the eccentricity circle $2e$,

$d_m = 2r$, $r = 2e$, and $d_m = 4e$.

Or, in terms of ratios,

$d_m/2r = 1$, $r/2e = 1$, $d_m/e = 4$.

Another independent parameter ratio used is P_s/e. Following the above logic, once the eccentricity e is obtained, the stator pitch P_s is calculated as eccentricity times this ratio P_s/e. The rotor pitch can then be calculated, as it is a function of the stator pitch and the lobe ratio. The remaining elements of the rotor and stator geometry can then be calculated, resulting in the cross-sectional fluid area, according to formula shown in Fig. 17. This will also produce the unit flow (gallons per revolution), which is directly related to the fluid area.

In summary, the choice of just one dimensional parameter (traditionally, this is a rotor minor diameter d_m, although any other dimension would work), and two key ratios (d_m/e and P_s/e) can uniquely define a pump's performance characteristics.

The choice of d_m/e and P_s/e defines pump (or mudmotor) geometries as *parametrically identical* designs for the same lobe ratio. This means that two designs, each having a different d_m but the same ratios (d_m/e and P_s/e), will appear identical on paper in a parametric (i.e., proportions) sense, as shown in Fig. 29. Thus, parametrically identical designs can have their geometry scaled up or down in direct proportion to their nominal size ratio (the ratio of their minor diameters).

This also explains why the rotational speed must be reduced for a large pump as compared to a small one. The wear rate affects the life of a progressing cavity pump, and this is a

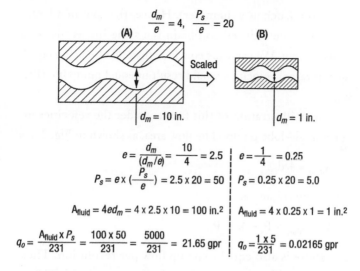

$$\frac{d_m}{e} = 4, \quad \frac{P_s}{e} = 20$$

(A)   Scaled   **(B)**

$d_m = 10$ in. $d_m = 1$ in.

$$e = \frac{d_m}{(d_m/e)} = \frac{10}{4} = 2.5 \quad \bigg| \quad e = \frac{1}{4} = 0.25$$

$$P_s = e \times (\frac{P_s}{e}) = 2.5 \times 20 = 50 \quad \bigg| \quad P_s = 0.25 \times 20 = 5.0$$

$$A_{fluid} = 4ed_m = 4 \times 2.5 \times 10 = 100 \text{ in.}^2 \quad \bigg| \quad A_{fluid} = 4 \times 0.25 \times 1 = 1 \text{ in.}^2$$

$$q_o = \frac{A_{fluid} \times P_s}{231} = \frac{100 \times 50}{231} = \frac{5000}{231} = 21.65 \text{ gpr} \quad \bigg| \quad q_o = \frac{1 \times 5}{231} = 0.02165 \text{ gpr}$$

- If a PC machine is scaled up or down x times, its areas scale as x^2, cavity as x^3, and thru-flow capacity as x^3. This is one of the reasons why smaller pumps tend to operate at somewhat higher speeds, to compensate for less of size effect.

FIGURE 29
Affinity relationships for PC machines.

function of the relative motion (rubbing) between the rotor and stator, and between the fluid passing through the internal passages. The latter is further exacerbated by the solids that are carried with the fluid, and the internal boundaries (the rotor and stator). Therefore, higher internal velocities reduce the useful life of a pump, and lower velocities increase it. Therefore, equal velocities would result in equal life. This is why a large pump of any type will wear out faster, since the velocities are proportional to a product of rpm times the linear dimension (such as a diameter). Hence, the rpm of a larger pump must be reduced to obtain the same life as that of a smaller pump. We can thus conclude that, all else being equal, equal internal velocities would result in equal pump life (Fig. 30).

For an illustration of this fact, consider the velocities inside a single-lobe pump. The flow area, as shown in Fig. 26, is

$$A_{fluid} = 4ed_m.$$

The cavity volume is

$$V = A_{fluid} \times P_s = 4ed_mP_s.$$

The above is also equal to pump flow per revolution. Thus, the flow is

$$Q = V \times rpm = 4ed_mP_s \times rpm.$$

The average fluid velocity (AFV) is equal to flow divided by the fluid area:

$$AFV = Q / A_{fluid} = P_s \times rpm.$$

So, for a constant wear rate (equal life), AFV must be constant (AFV = const):

$$rpm = const / P_s.$$

For parametrically equal designs (having the same ratios d_m/e and P_s/e),

$$P_s = e \times (P_s/e) = d_m / (d_m/e) \times (P_s/e) = d_m \times const,$$

or

$$rpm = const / d_m.$$

Similar conclusions can be drawn by analyzing the equations for the rubbing velocity and maximum particle velocities, which were shown in Figs. 27 and 28.

The above shows that the rotational speed (rpm) must be reduced in direct proportion to the pump size to achieve the same pump life (i.e., a direct effect on the equipment reliability and MTBF).

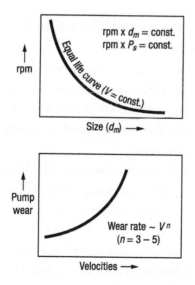

FIGURE 30
Equal velocities mean equal wear rate, but the wear rate increases exponentially with velocities. This is why larger pumps need to run slower.

It should be noted here that, while equal velocities result in equal life, *nonequal* velocities do not necessarily result in a *linear* relationship between the velocities and pump life. In fact, the exact relationships between the equipment wear rate and velocities are complex, comprising an entirely separate subject on material wear phenomena. The relationship between the wear rate and fluid (as well as rubbing) velocity is usually exponential:

Wear rate ~ Velocity^n.

The exponent n varies, typically n = 3–5 (see Fig. 30).

Variation of Ratios and Their Effect on Performance and Life

The discussion above related the performance and reliability (via wear rate) for different-sized pumps having the same key ratios. It was shown that to achieve the same wear rate in different pump sizes, the rotating speed of *parametrically similar* pumps must be changed.

However, if the key ratios are varied for basically the same pump sizes, it is possible to change other performance factors, such as to increase pump life; increase or decrease through-flow for a given rpm (for example, if the same motor drive is to be utilized); modify torque capabilities (a critical aspect for mudmotor machines); reduce vibrations; or improve self-priming ability.

Some of these goals can be conflicting, and improvement in one area may have a negative effect in another. Factors that influence how key ratios affect different aspects of pump operation are listed below.

Diameter-eccentricity ratio (d_m/e): The same *throughput flow capability* (in terms of unit flow, or the same flow at the same rpm) can be accomplished using an infinite number of different d_m/e ratios (Fig. 31), assuming the same pitch, same tube size, and minimum elastomer thickness. This is possible if the net fluid area is kept constant while the ratio (d_m/e) is the same.

This has an interesting implication for the effect on pump vibrations. Progressing cavity machines are inherently unbalanced because their rotors are not straight; they nutate around the stator center, offset by the eccentricity. The centrifugal force that causes vibration is equal to the mass (or weight) of the rotor times the rotational speed squared, times the eccentricity:

$$F \sim W \times rpm^2 \times e \sim A_{rotor} \times P_r \times rpm^2 \times e \sim (r^2 \times e).$$

Since we are comparing rotors having equal pitches, the weight of the section along the rotor pitch distance can be considered for comparison. This weight is equal to the rotor cross-sectional area times the pitch. The cross-sectional area (function of the r^2 or $d_m{}^2$) of the rotor in Fig. 31 (Case B) is greater then that for Case A. The eccentricity, on the other hand, is greater for Case B. The product ($r^2 \times e$) is thus a comparable measure of centrifugal force and, as a result, of pump vibrations.

As it turns out, the added contribution from the r^2 is greater than the reducing effect of the smaller e, which results in $r^2 \times e$

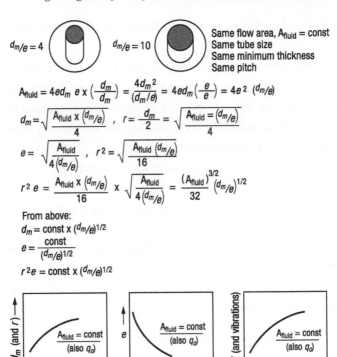

$d_{m/e} = 4$ $d_{m/e} = 10$ Same flow area, A_{fluid} = const
Same tube size
Same minimum thickness
Same pitch

$$A_{fluid} = 4ed_m \ e \times \left(\frac{d_m}{d_m}\right) = \frac{4d_m^2}{(d_m/e)} = 4ed_m\left(\frac{e}{e}\right) = 4e^2 \ (d_{m/e})$$

$$d_m = \sqrt{\frac{A_{fluid} \times (d_{m/e})}{4}} \ , \ \ r = \frac{d_m}{2} = \sqrt{\frac{A_{fluid} = (d_{m/e})}{4}}$$

$$e = \sqrt{\frac{A_{fluid}}{4(d_{m/e})}} \ , \ \ r^2 = \sqrt{\frac{A_{fluid} \ (d_{m/e})}{16}}$$

$$r^2 e = \frac{A_{fluid} \times (d_{m/e})}{16} \times \sqrt{\frac{A_{fluid}}{4(d_{m/e})}} = \frac{(A_{fluid})^{3/2}}{32}(d_{m/e})^{1/2}$$

From above:
$$d_m = const \times (d_{m/e})^{1/2}$$
$$e = \frac{const}{(d_{m/e})^{1/2}}$$
$$r^2 e = const \times (d_{m/e})^{1/2}$$

FIGURE 31

Force and vibrations are greater with an increased d_m/e ratio if the tube ID remains the same. However, less rubbing would result in longer elastomer life for a higher ratio of d_m/e.

being greater for the rotor with less eccentricity. In other words, for a given net fluid area, the unbalanced force is greater for a larger d_m/e ratio. If the d_m/e ratio is varied while d_m is kept the same, then the unit flow (*throughput flow capability*) is decreased for higher ratios, as can be see from Fig. 32. However, through-

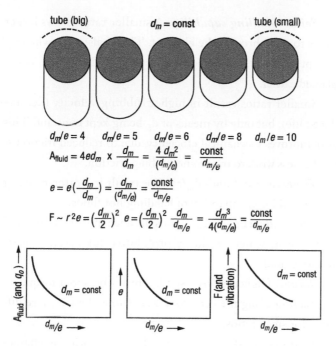

FIGURE 32
Force and vibrations are reduced with increased d_m/e ratios, while the tube size can also be reduced.

flow capability is also reduced. Elastomer life improves at larger d_m/e ratios due to reduced rubbing velocity.

In this case, the centrifugal force is greater with lower d_m/e ratios. Of course, in this case, either the thickness of the elastomer or the tube size must change to accommodate a smaller d_m/e ratio, for a constant d_m dimension. A smaller d_m/e ratio with a constant d_m results in thicker elastomer at the minor diameter. This may cause the stator to run hot, reducing its life.

Solids handling capability: A smaller ratio allows larger solids to pass through the element. This is also true when d_m is kept constant, resulting in a larger net fluid area, as explained earlier.

Smaller ratios result in higher rubbing velocity (i.e., reduced life), but only by means of d_m being kept constant. This also improves mechanical efficiency, due to reduced friction—it takes less work to overcome rubbing friction.

Temperature: Small d_m/e ratios result in hotter-running stators and lower capability of handling hot fluids.

Shear rate: Higher d_m/e ratios produce more agitation and more shear, which could be prohibitive for certain shear-sensitive fluids.

Torque: Higher ratios result in greater breakout torque.

Sealing along the rotor/stator seal-line: This is more difficult for small ratios.

Priming: Small ratios can result in poor priming characteristics.

All of the above indicates that higher ratios are more desirable. The only downside is that the higher ratios result in larger (more expensive) pump sizes. As a general rule, however, a somewhat greater initial investment in a larger unit can result in significant operating savings in parts, labor, and reduced downtime.

Pitch-eccentricity ratio (P_s/e): Varying the pitch-eccentricity ratio will have the following effects.

Wear rate: The higher the ratio, the higher the wear rates in the pump element.

Helix (scroll) shape: Higher ratios produce a more gradual ("lazy") helix (Figs. 33 and 34).

Throughput: Higher ratios result in higher flow.

Cost: A long-pitch pump is easier to manufacture for a given flow rate.

NPSHr (net positive suction head required): Larger ratios have poorer suction characteristics.

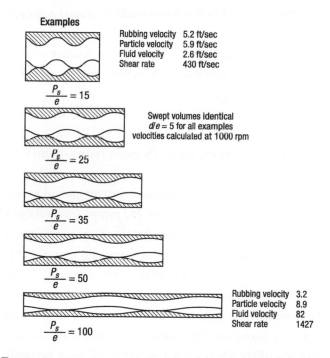

Examples

Rubbing velocity	5.2 ft/sec
Particle velocity	5.9 ft/sec
Fluid velocity	2.6 ft/sec
Shear rate	430 ft/sec

$\dfrac{P_s}{e} = 15$

Swept volumes identical
$d/e = 5$ for all examples
velocities calculated at 1000 rpm

$\dfrac{P_s}{e} = 25$

$\dfrac{P_s}{e} = 35$

$\dfrac{P_s}{e} = 50$

$\dfrac{P_s}{e} = 100$

Rubbing velocity	3.2
Particle velocity	8.9
Fluid velocity	82
Shear rate	1427

FIGURE 33
Longer pitch increases particle and fluid velocity, which leads to static pressure reduction (i.e., NPSHr is better with a shorter pitch).

Viscous fluids: It is easier to handle viscous product with lower ratios because the shear rate is high at high ratios and lower (better) with lower ratios.

AFV: Higher ratios result in higher average fluid velocity.

Particle velocity: Higher at higher ratios.

Product damage: More likely at higher ratios.

Seal line: Wider at higher ratios, and more difficult to seal (i.e., lower pressure capability per stage).

Starting torque: Higher ratios result in greater required starting torque.

Particle jamming: High ratios result in smaller contact angles between rotor and stator, which increases the possibility of particles being trapped.

As can be seen from the above list, the choice of the pitch ratio is less straightforward as compared to a diameter ratio, as many of the effects caused by the pitch ratio variation are conflicting.

The design approach to progressing cavity pumps versus mudmotors is significantly different. PC pumps are always of a single-lobe design, with two-lobed rotors as rare exceptions. Mudmotors, on the other hand, are essentially always multi-lobe designs to maximize torque per unit of length. These dimensional constraints are dictated by the nature of their application in a drilling hole, where diameter and length must be kept to a minimum. This is especially true for angular and horizontal drilling where drill-strings (Fig. 35) must negotiate small radii of curvature. As a result, dimensional and power

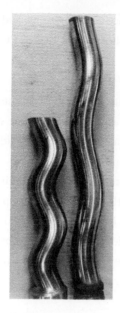

FIGURE 34
Rotors with various pitches.

FIGURE 35
A mudmotor power section.

requirements override endurance considerations, and the over-all life of mudmotors is much shorter than that of PC pumps.

Power Transmission Methods

A design of a progressing cavity pump (Fig. 36) must be able to transfer a simple concentric rotary motion of the shaft of an electric motor to a pump rotor, which nutates inside the stator, as was explained in detail in Chapter 4.

To accomplish this, an intermediate shaft (called a connecting rod), which connects the rotor to the pump drive end shaft, is employed. Two joints (one at each end) participate in rotational and oscillatory motion, allowing the rotor to nutate around the eccentricity *e*. Several designs are available to accomplish this. The most traditional design utilizes a *pin joint*, as shown in Fig. 37.

A pin joint is simple and relatively inexpensive. The cavity inside the joint must be sealed by a lip seal or a rubber boot to prevent pumpage from entering it and jamming the joint. It is

FIGURE 36
Typical progressing cavity pump. Concentric motion of the driver shaft is transmitted to a nutating rotor via a connecting rod with two joints at each end.

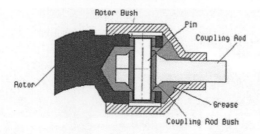

FIGURE 37
Pin joint detail.

also filled with grease to lubricate the interface between the bushing of the connecting rod and the pin. The bushing and pin are usually made from hardened steel, but they eventually wear out and must be replaced. Figures 38 and 39 show closer views of the pin joint and a connecting rod that attaches to the rotor and the drive shaft at each end.

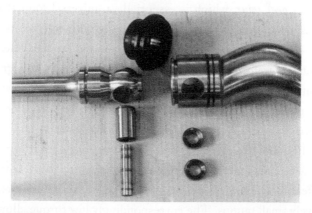

FIGURE 38
Pin joint part details.

FIGURE 39
Connecting rod attaches to the rotor and drive shaft via pin joints.

A somewhat more expensive but more robust design utilizes a *gear joint*, which is essentially a universal joint that transmits torque via gear teeth and allows continuous rotation and oscillation. It is also filled with grease and is sealed. Gear joints can transmit higher torque as compared to a simple pin joint, and typically last several times longer, which often justifies their expense for critical applications where downtime is especially costly. A gear joint is shown in Figs. 40 and 41.

In many installations, there are situations where a progressing cavity pump must move fluid against relatively small differential pressure, such as in transfer applications using relatively small pumps. The correspondingly low torque allows one to eliminate the joints altogether; instead, a flexible connecting shaft is used, as shown in Fig. 42.

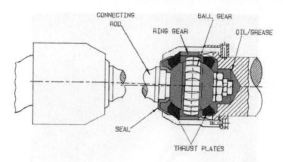

FIGURE 40
Gear joint.

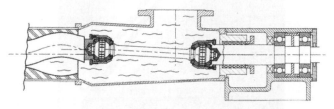

FIGURE 41
Gear joint connection.

This connecting shaft (or connecting rod) is usually made from a high-tensile-strength nonmetallic material, and its kinematics are illustrated in Fig. 43. These designs became popular in the mid-1990s, due to their simplicity, low cost, and ease of maintenance.

Starting with smaller units, flexible-shaft designs have further improved and are now used even for larger units (Fig. 44).

While pin and gear joints require that their ends be sealed, flexible joints do not need seals, which improves pump ability

FIGURE 42
Flexible shaft eliminates joints.

to handle stringy, fibrous materials and foreign entrapments, such as rags (Fig. 45).

Pump Sealing: Traditionally, PC pumps have been used in viscous, dirty, and solids-laden applications. However, from the chemical view, these materials are usually not exceptionally aggressive, and zero-leak demand is not usually required. Traditional *packed-box* sealing arrangements have been successfully employed for PC pumps and are still a common seal-

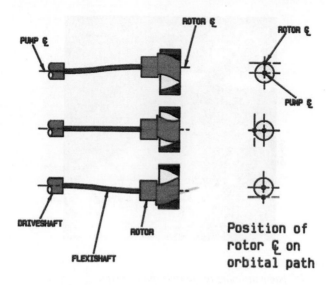

(Note: for clarity flexishaft deflection exaggerated)

FIGURE 43
Principle of operation of a flexible shaft.

ing option; such arrangements are no different than that required for any other pump type. Figure 46 (top view) shows six packing rings stuffed inside the box by the gland. Some leakage is required to lubricate packing rings, which otherwise would run dry and burn, damaging the shaft. A typical leakage rate is 10–20 drops per minute. The packing box can be equipped with a lantern ring, as shown at the bottom portion of Fig. 46.

This allows external injection (typically, of water) at pressure higher than the pump inlet pressure. A portion of the injected water gets inside the pump, and some leaks outside, but

FIGURE 44
Large PC pump utilizing the flexible shaft concept.

the outside leakage is much cleaner than if no injection was used. If the pumpage contains chemicals and leakage to the outside cannot be tolerated, a *mechanical seal* is used. Mechanical seals have no appreciable leakage rate and their emissions are measured in parts per million, typically 50–500 PPM. However, ordinary, nonflushed, single mechanical seals can get easily plugged, since the pumped material is usually dirty or too viscous and does not provide sufficient cooling and lubrication of the seal faces (as is often the case with PC pumps). Such pumps, if starved for lubricating liquid, would burn and fail. A double mechanical seal with a barrier fluid injected between the primary and secondary seals can improve the lubrication process near the seal faces, and can appreciably extend their life. Instead of standard double mechanical seals, however, PC pumps use a

FIGURE 45
Flexible shaft design allows pumps to improve handling of stringy, fibrous material, such as rags, without clogging the joints.

COMPONENT - SHAFT SEALING

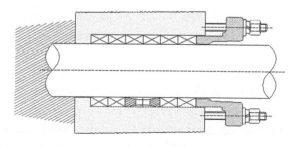

Packed Gland

FIGURE 46
Packed-box sealing of a pump.

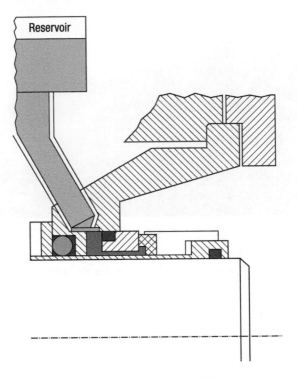

FIGURE 47
Buffer fluid keeps the seal cavity clean, extending life.

combination of a single mechanical seal as a secondary seal, and a lip seal facing the pumpage (Fig. 47).

This removes the mechanical seal from close proximity to the pumpage and simplifies the design. A tradeoff is wear of the lip seal, which typically lasts about 2,000 hours for a pump at full speed (1,800 rpm). However, since PC pumps operate at much lower speeds, their lip seals last longer.

Chapter 6

APPLICATION GUIDELINES

Like any other pump type, PC pumps have their strong points as well as limitations. The Hydraulic Institute chart, shown at the beginning of this book, depicts thirty-three different types of pumps that are grouped by their design similarities, under two basic classes: centrifugal and positive displacement pumps. Given this variety, how does, or should, a user make a decision to select and apply a particular pump type?

In most cases, the choice is simply a preference based on historical tradition. If a particular pump type has worked well in the past, why change? Reliability and the production schedules of operating plants are critical, and experimenting with something new, not proven at a plant, is not a decision taken lightly. Such user conservatism in applying new types of pumps at the existing facilities is understandable. Only when significant process changes at the plant are taking place, or if the reliability of the present pumping equipment becomes, for whatever reason, significantly lower than the historical plant average, do maintenance and engineering personnel consider switching to a different pump type.

For new installations, however, the selection of pumping equipment can be less conservative. While recommendations from the maintenance and engineering departments from the existing operating plants are considered, the decision on the type of equipment for a new plant is usually more open-minded, and open bidding allows manufacturers of different types of pumps to present their offerings. Price plays a significant role, of course, but not until the technical aspects of the equipment are confirmed to be in compliance with the operating requirements. In this respect, reliability and dependability of the equipment are critical because plant downtime costs many times more than the initial purchase price of the pump(s). Only if all other operating aspects are equal, the pricing considerations enter the picture for a final decision.

As a general rule, progressing cavity pumps are selected for hard to pump, very viscous, dirty, but not extremely corrosive fluids. Examples are slurries, latex, sewage, resins, varnish, starch, paper pulp, asphalt, paste, paints, mud, cement, and the like. Exceptions are common, however, and it is not unusual to encounter applications where progressing cavity pumps pump kerosene, gasoline, water, etc. For such cases, the pressures are relatively low, temperatures are around ambient, and the liquids are not aggressive. Such exceptions are often made based on logistics rather than necessity. For example, a plant might have had surplus inventory of pumps for other applications and, when a need arose for a different installation where a more traditional (e.g., centrifugal) pump would be normally used but it was not readily available in stock, the maintenance department might have installed a less common pump type for that application, on an emergency basis. If the temporary pump proved to work fine, that could have led to a decision to keep it there. Then, if a manufacturer's representative of the new temporary pump demonstrated fast response and good follow-up support, the pump might become "cemented" for the application, even though the pump itself might not have been the absolute optimum from other aspects. The bottom line is if it works and price is *reasonable*, use it!

Abrasion

Progressing cavity pumps are hard to beat when it comes to abrasive fluids. A diaphragm pump is another choice, and both have the advantage of requiring much less floor space. However, diaphragm pumps produce significant pulsations that propagate through the piping, resulting in piping and foundation structure vibrations. Pulsation dampeners are almost always required with diaphragm pumps, and their failure, clogging, or incorrect positioning could result in system-wide problems. Progressing cavity pumps, on the other hand, are practically pulseless, resulting in smooth, quiet operation.

With PC pumps, the size of the pumpage particles is limited only by the fluid cross-sectional area, which is easy to determine by just looking at the open end of a pump. For a given required flow rate, a larger pump operating at a lower rpm would handle larger solids, with less wear, due to its lower speed. Many factors affect the wear rate but, as a rough rule of thumb, the *life of a PC pump can be assumed to be the cube of the rpm.*

Temperature

A critical element of a PC pump is the stator elastomer lining. Most elastomers cannot be used in very high temperature applications; most PC pump applications are under 180°F. The most widely used elastomer for stator linings is rubber, such as nitrile or natural EPDM. Teflon® elastomers are also used, although less frequently, and can be applied where temperatures reach approximately 300°F.

Chemicals

Progressing cavity pumps do not handle aggressive chemicals well. Substances like sulfuric acid, caustics, and the like, are best left to other pump types, such as stainless gear pumps (Fig. 48), for relatively low flow rates, or centrifugal pumps for moderate to high flow rates.

Viscosity

Progressing cavity pumps can move substances up to 1,000,000 cP (Fig. 49), which is beyond the range of most other pump types.

The major challenge in pumping such highly viscous substances is the ability of the substance (which hardly seems to be a fluid!) to get to and enter the pump inlet. Special augers and wide-throat hoppers are used in such instances, to assist the fluid entering the pump and to prevent bridging. An example of such a design is shown in Fig. 50.

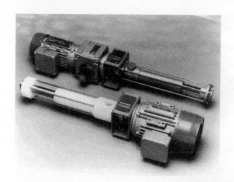

FIGURE 48
Comparison between typical applications of rotary gear pumps (bottom, courtesy of Zenith Pump) and progressing cavity pumps (top, courtesy of Monoflo Company).

FIGURE 49
PC pumps can pump extremely viscous substances.

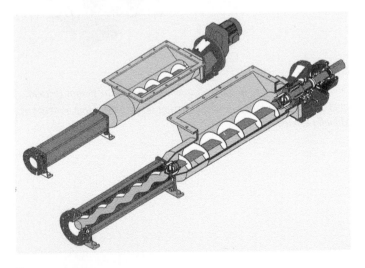

FIGURE 50
This design option has a wide-throat hopper with an auger-assisted bridge-breaker, to handle very viscous products (courtesy of Monoflo Company).

Speed (rpm)

Even though progressing cavity pumps can, theoretically, run as fast as other pump types (1,800 rpm), they rarely do. The first reason is that the very geometry of nutating rotors results in unbalance (i.e., centrifugal force due to rotor mass offset [eccentricity] from the pump centerline). This force, according to Newton's law is

$$F_{centrif\,(unbalance)} \sim m \times rpm^2 \times e.$$

This means that the faster the pump runs, the greater the unbalance force, and thus the greater the vibrations.

The second limitation on speed is viscosity. If the fluid enters the pump inlet successfully, the pump internal element (i.e., rotor) takes over the task of moving the fluid on and the product will come out at the other end. The flow of any positive displacement pump (including progressing cavity pumps) is constant for a given rpm, and is practically independent of the viscosity. Note, however, that at very low viscosities (under ~ 30–100 cP) a slip influence on net flow is appreciable. What changes is the required power, which increases to turn the rotor to displace a more viscous fluid. Until structural or driver horsepower rating limitations are reached, the pump will continue pumping normally. However, the flow of fluid from the supply tank to the pump inlet can only occur if there is sufficient pressure on the inlet side, which pushes the fluid to enter the pump cavity where a pumping action commences.

If a pump turns too fast, it may be pumping more fluid through itself as compared to the ability of the inlet pressure to deliver the same amount of fluid to the pump inlet cavity opening. If this happens, the inlet cavity will be incompletely filled, resulting in cavitation, vibrations, noise, and the loss of flow. According to Hydraulic Institute data, a minimum required suction pressure is defined as the pressure in front of the pump suction below which a positive displacement pump exhibits a loss of more than 55% of its flow. Pumps should not operate below the minimum required suction pressure; preferably, some margin above that value should exist. A practical implication for pumping system design is to reduce the inlet losses as much as possible with shorter pipe runs between the supply tank and the pump; larger pipes; and fewer elbows, turns, and bends, etc.

A third limitation on speed is solids content. For example, progressing cavity pumps handle coal slurries where the carrier fluid is water (i.e., not a viscous substance). However, coal particles can damage the rubber lining if they move too fast within the internal passages of the pump. Thus, slowing the pump rpm can significantly increase the life of the lining.

One or all of the above factors are usually present during PC pump operation: unbalance (always present), high viscosity, and particle size. As a result, PC pumps usually do not run at speeds above 300–500 rpm, and often even lower. For very viscous fluids, it is not uncommon to have progressing cavity pumps operating at 10–50 rpm.

Pressure and Flow

Progressing cavity pumps are known to pump flows over 2,000 gpm (Fig. 51).

FIGURE 51
This large PC pump moves over 200 gpm, at about 70 psi differential pressure, with a 200-HP motor and a gear reducer, allowing the pump to run at 150 rpm.

Usually, pressures are low at high flows, and several pumps can often be installed in parallel to minimize the size of each individual unit, due to manufacturing considerations. Higher pressure means more slip, which, even at tighter rubber-to-rotor fits, may reduce the net throughflow significantly. This is why the higher the differential pressure, the more stages are required (Fig. 52).

As a rule of thumb, progressing cavity pumps are designed for approximately 75 psi differential pressure per stage. For example, a 300-psi application would require 4–5 stages (staging error should be on a plus side). For a preliminary understanding

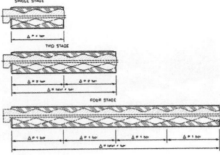

FIGURE 52
High differential pressure requires more stages.

of the overall pump dimensions, consider the following example in which a progressing cavity pump is required to move 300 gpm against 300 psi differential pressure.

When sizing a pump, design engineers use formulas and charts referencing a variety of theoretical and empirical data. For example, as discussed in Chapter 2, rubbing velocity and particle maximum velocity are the primary limiting factors on pump speed. However, a pump user typically does not have access to this specialized information. Even a relatively simple criterion, the average fluid velocity (AFV), as discussed in Chapter 5, requires knowing the net fluid area, which can be measured only if the user already has an operating pump. Obviously, during the pump evaluation and selection process, most users do not yet have a pump.

Using a pump catalogue is a convenient way to estimate the required pump size if it shows comparative performance curves (flow versus pressure charts for several pump sizes). However, because catalogues are not always available, it would be good to have some simple, easy method to estimate the overall pump dimensions. It turns out that such an approximation is possible if one knows only the nominal dimension of the connective piping. This information is usually available, regardless of the pump type. Even though the pipe size varies according to the fluid viscosity (more viscous fluids require larger pipes), such variations are relatively stable and known, especially if used only for rough guidelines.

The empirical rule that can be used for approximating pump size is 1–3 ft/sec fluid velocity, based on the nominal pipe size. A smaller number (1 ft/sec) is usually used for smaller pumps, and the larger (3 ft/sec) for larger sizes:

$$V_{pipe} = Q \times 0.321 / A_{flange\ nominal}$$
$$= Q \times 0.321 / (3.14 \times d_{flange}^2 / 4).$$

In the example above (300 gpm, 300 psi), let's assume 2 ft/sec for this average-size pump. If the above equation is revised to solve for the flange (i.e., pump size) diameter, we get

$$d_{flange} = (4 \times Q \times 0.321 / (3.14 \times V))^{0.5}$$
$$= (4 \times 300 \times 0.321 / (3.14 \times 2))^{0.5} = 7.8 \text{ in.}$$

A nominal 8-in. pipe (pump flange) can thus be assumed.

The total length of the hydraulic section (rotor inside the stator) can vary but, as a first estimate, take the length of one stage equal to roughly four nominal diameters:

$$L_{1\text{-stage}} \sim 4 \times d_{flange} = 4 \times 8 = 32 \text{ in.}$$

Using five stages (according to 300 psi, and 75 psi per stage, plus a margin):

$$L_{hydraulic} = 5 \times 32 = 160 \text{ in.}$$

The drive end adds more length, and is probably no less than at least one or two stages as in

$$L_{pump} = 160 + 32 \times 2 = 220 \text{ to } 230 \text{ in.} = \sim 20 \text{ ft!}$$

To minimize floor space, PC pumps often have "piggy-back" mounted motors with belt drives, as shown in Fig. 53.

Applying the same logic to a smaller pump, such as 20 gpm and the same 300 psi differential pressure:

$$d_{flange} = (4 \times 20 \times 0.321) / (3.14 \times 1))^{0.5}$$
$$= 2.9 \text{ in.} \sim 3 \text{ in. (using 1 ft/sec)};$$
$$L_{1\text{-stage}} \sim 4 \times 3 = 12 \text{ in.};$$
$$L_{hydraulic} = 5 \times 12 = 60 \text{ in.; and}$$
$$L_{pump} = 60 + 12 \times 2 = 84 \text{ in., or} \sim 7 \text{ ft.}$$

Granted, the assumptions above are very approximate, but this example illustrates the point that a typical PC pump, when mounted on a base plate with the motor, may require about twice the floor space as compared to most other pump types. Furthermore, if the abrasives content is significant, the differential pressure rating of 75 psi should be reduced even further, thus making the overall pump even longer.

This explains why it is often difficult to replace other, even problematic, pump types at existing installations (where the surrounding floor space is usually crammed with equipment). However, for outdoor installations and, especially, for new

FIGURE 53
"Piggy-back" mounted motors are often used with PC pumps to save floor space.

installations, this is less of an issue because the floor plans are just being established and accommodations for a longer pump can be easily accounted for.

Entrained Gas

Multiphase flow presents application opportunities for PC pumps. Oil recovery often requires pumps to handle the entrained and/or dissolved gas. Such applications are challenging for several reasons. First, the mixture is not stable; it alternates among being mostly liquid, liquid-gas, and mostly gas. If a mostly gas mode continues for an extended time, no lubrication is available for the rotor-to-stator interface, which would result in rapid failure. Temperatures can exceed 300°F and render the elastomer unsuitable. In addition, as the mixture of liquid and gas progresses through the pump from lower pressure to higher, more gas dissolves and the net volume is reduced, resulting in the incomplete filling of the cavities, creating associated noise and vibrations.

Nevertheless, these challenges are equally relevant to all pump types, not only PC units. The ability of PC pumps to handle unusual mixtures is a definite advantage and, once new technologies and engineering enhancements are further researched, PC pumps might become a viable option for multiphase oil and gas recovery applications (Fig. 54).

Entrained Gases & Froth

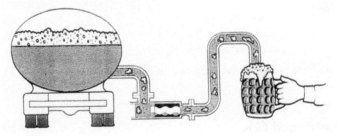

FIGURE 54
Multiphase pumping by PC pumps.

Dry-Running

PC pumps should not be run dry because the elastomer, rubbing against a dry interface with the rotor, can quickly fail. While this appears to be in contradiction to the otherwise excellent self-priming ability of PC pumps, the fact is that PC pumps will self-prime and operate well if restarted after a relatively short period of being idle. In these cases, some residual liquid is still present inside the pump, providing limited but sufficient lubrication during the self-priming cycle, which usually does not take more than 30–60 seconds. It is critical that when PC pumps are started for the first time or restarted after a prolonged idle period, they should be manually primed prior to startup; otherwise, such "bone-dry" operation can

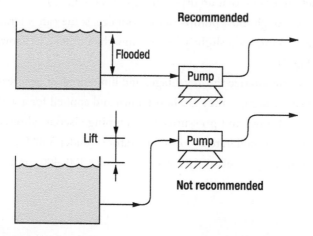

Figure 55
If at all possible, it is better to keep the pump inlet flooded at startup.

FIGURE 56
Limited dry-running capability can be achieved with special designs incorporating components made from engineered composite materials that include graphite, which has good self-lubricating qualities (courtesy Sims Pump).

result in quick failure of the elastomer. For this reason, it is always recommended for any pump type that the pump inlet be flooded (having the liquid supply level above the pump inlet), to ensure that no dry-running occurs (Fig. 55).

Although most pumps cannot run dry, some can tolerate dry-running for a slightly longer time than others, as shown in Fig. 56.

To summarize the advantages and limitations of progressing cavity pumps: they are best known and applied for a wide range of flows and pressures; for pumping viscous, abrasive products at moderate to low temperatures (under 300°F); and for operating at relatively low speeds (under 300–400 rpm).

Chapter 7

INSTALLATION EXAMPLES

Installation Case 1: A PC Pump Maintains Foam Concentration

The West Midlands Fire Service Co. (WMFS) in the U.K. needed a way to pump liquid protein concentrate from storage containers at the fire station into 500-gallon foam tender units, ready to be used on chemical and gasoline fires (Fig. 57). Although this fire service has considerable experience with centrifugal pumps, these are not suitable for pumping the foam concentrate. They create too much aeration, resulting in the foam tender units being just 25% full of concentrate and 75% full of foam, totally unacceptable for an effective service.

To fill a foam tender unit completely without a pump required the manual effort of five men over several hours, providing there were no interruptions (such as being called out on an emergency). There are three foam tender units strategically placed throughout the area covered by the WMFS, near highways,

Figure 57
Installation example 1 (courtesy Monoflo Pumps Co.).

airports, and large chemical installations. The vulnerability of manually loading the foam tenders was highlighted when a gasoline tanker truck crashed off a major freeway, spilling gasoline down an embankment, which ignited shortly after the first fire appliance arrived. With two special foam tender units already mobilized, the fire chiefs realized that with only one unit on standby, and with at least three hours needed until the other two could provide total foam tender cover, the region was vulnerable should there be another major incident.

Extensive trials with centrifugal pumps proved fruitless. Accordingly, a progressing cavity pump manufacturer was given a list of requirements that its PC pumps would have to fulfill. These were: no aeration of liquid, simple operation, high reliability, fail-safe operation, and long life.

After successful trials, WMFS placed an order for three SE061 progressing cavity pumps, fitted with stop/start reversible controls and over-pressure sensing switches. The manufacturer also suggested that the units be fitted with timers and a special "snorer bypass" for dry-run protection. The pump's ability to pump in both directions could be used to empty the foam units for maintenance purposes.

The benefits of this pump type were immediate. It now takes two people just twelve minutes to fill a foam tender unit—a task that previously took up to four hours. Using the SE061 model PC pumps has thus contributed to a significant increase in the efficiency of the foam tender units.

Installation Case 2: A Sludge Pumping Problem

Following successful trials, close-coupled progressing cavity pumps have been installed at the Thames Water's Slough (U.K.) sewage treatment plant, replacing centrifugal pumps on macerated sludge duties (Fig. 58). This was part of a sludge digester and macerating equipment refurbishment project contracted by Environmental Construction Ltd. of Salisbury, U.K.

One centrifugal pump was originally replaced with a PC pump coupled to a fixed-speed gear box to pump macerated sewage sludge with a solids concentration of approximately 5%. Following a successful three-month trial period, the two remaining centrifugal pumps on this duty were replaced with PC units. One significant benefit was their compact design, which makes them ideal for installation where space is at a premium. PC design has the ability to handle liquids of various

Figure 58
Installation example 2 (courtesy Monoflo Pumps Co.).

viscosities, from water to slurries and dewatered sludges. Such pumps will also handle fluids containing fibrous, stringy solids, making them particularly suitable for the water and waste water treatment industries.

In this installation, the rotor drive coupling uses a sealed pin joint developed from a design proven on many thousands of applications. The angle of articulation, caused by the eccentricity and length of the coupling rod, is lower than that found in many other PC pumps with short coupling rods. This specific design offers significantly reduced pin joint wear compared with other pumps running at the same speed. The pumps installed at the Thames Water's Slough sewage treatment plant run at 208 rpm.

Reliability is further enhanced by the fact that the manufacturer makes its own rotors and molded-to-metal stators, resulting in total control over tolerances during manufacture, thus ensuring consistent performance. The gear box is flange-mounted, with a plug-in shaft for ease of dismantling and assembly during routine maintenance. Since being installed at this facility in November 1992, the pumps have continued to operate without problems.

Installation Case 3: A Food Application— Soya Milk Processing

Progressing cavity pumps were installed at the Haldane Foods Soya Milk production facility in Cheshire, U.K., for the efficient movement of material with a high solids content (Fig. 59).

The quality and taste of the finished product depends upon careful processing of the finest soya beans to preserve their nutritional value; this is achieved by maintaining accurate control during all stages of the manufacturing process.

After careful deactivation of the enzymes, the beans are ground with water to form a consistent, thick paste, and a PC pump is used to pump this to a separator. The smooth-flowing characteristics of PC pumps are utilized here to good effect, providing a constant feed rate to decanter centrifuges and avoiding surges to ensure that the paste has a smooth, even consistency. The pumps installed at Haldane are required to handle constant operating temperatures of up to 194°F. After separation, the extracted liquid undergoes a vacuum deodorization before being processed and packed. It is sold as a milk substitute and used in a range of products such as nondairy ice cream, yogurt, and cream.

FIGURE 59
Installation example 3 (courtesy Monoflo Pumps Co.).

At this plant, removing the fibrous residue had been a slow and labor-intensive manual process. The installation of a PC unit to pump this effluent away into containers through a 3-in. pipe proved to be an excellent solution with a rapid payback period. The PC pumps were installed as part of a major process modernization and upgrading program that was carried out over a period of six months. Haldane Foods can now produce soya milk at the rate of 800 U.S. gallons per hour.

Installation Case 4: A Sanitary Application at a Bakery

PC pumps were effectively applied as part of a new product transfer system at Manor Bakeries in Manchester, U.K. (Fig. 60).

To allow cost-effective bulk deliveries of preserves to be accepted, the production team at the bakery investigated the feasibility of installing pipework and the associated equipment to automate the transfer of preserves to a new pie-filling machine. The resulting design for the new system included 200 feet of pipework with eight bends and 16 feet of static head.

FIGURE 60
Installation example 4 (courtesy Monoflo Pumps Co.).

Samples of the preserves were sent to the pump manufacturer's laboratory, where the viscosity at various shear rates was measured in order to calculate the system backpressure and thus determine the optimum pump size. These sanitary pumps were found to be ideal for the duty and were installed as part of the new cost-saving system.

A key feature of this particular design variation links the bearing shaft to the helical rotor. This device, developed by the pump manufacturer in the late 1960s, is ideal for sanitary applications; since it has no moving parts, there is no wear and therefore lubrication is unnecessary. Product contamination is eliminated. The simplicity of this joint design also reduces the number of parts in the pumps, making them easy to dismantle, service, and reassemble. Increased reliability also lengthens the intervals between routine maintenance.

These pumps were mounted on stainless steel base plates to facilitate cleaning, and fitted with mechanical seals and variable speed drives. The system has operated successfully since the mid-1980s.

Installation Case 5: Effluent Treatment

Progressing cavity pumps have provided reliable service since their 1996 installation at an effluent treatment project implemented by the oil-refining company Eastham Refineries of Ellesmere Port, U.K. (Fig. 61).

In 1992, Eastham completed its effluent treatment project, which was planned and implemented entirely by Eastham's own personnel. In this system, unwanted water from various pro-

FIGURE 61
Installation example 5 (courtesy Monoflo Pumps Co.).

cesses, combined with runoff surface water, is channeled to a large interceptor pit where oil and other chemicals are skimmed off and safely disposed of. The remaining water is then pumped by these PC pumps to a holding tank where it is again skimmed, making it ready for biotreatment (decontamination, reoxygenation) and eventual release to the water network.

The two PC pumps supplied are direct-coupled to gear boxes with explosion-proof motors. They provide respective pumping capacities of 176 and 264 U.S. gallons per minute. Both pumps are used on duties where the liquid consists of water with traces of oil and grit and the pumping temperature is 68°F. The pumps maintain a suction lift of 13–16 ft and operate with a differential pressure of 58 psi.

The choice of these PC pumps to service the system's two holding tanks virtually made itself. A number of other PC design variations from the same manufacturer were already installed around the plant, and they had all provided efficient, dependable service. Due to its pumps' reliability and ease of operation, this pump manufacturer was the first choice of operating and maintenance personnel.

The nature of the project meant that Eastham required extremely reliable, competitively priced equipment. The project engineering manager stated, "We purchase the best equipment for the job, build in standby facilities, and expect the highest performance standards from it. These PC pumps are in stainless steel with nitrile rubber stators and benefit from a very simple design which is easy to repair. If the pumps have needed attention, we have found [the manufacturer] operates a very efficient spare parts service."

Such all-inclusive and efficient service is of utmost importance to the pump users.

Installation Case 6: A Storm Water Runoff Application

After carefully considering a variety of options for handling storm water at its West Bromwich, U.K., resin manufacturing plant, Hepworth Minerals and Chemicals, Ltd. chose progressing cavity pumps to provide the solution (Fig. 62).

The system was designed to meet the extremely stringent requirements for the discharge of storm water. Eight pumps were installed to transfer storm water runoff from sumps to a

FIGURE 62
Installation example 6 (courtesy Monoflo Pumps Co.).

holding tank, prior to treatment and discharge. PC pumps also handled the discharge. These pumps were chosen because they provide a host of long-term benefits. They are surface mounted, which provides easy access for pump and motor maintenance.

The wide variety of seal options available reduces the chances of a seal failure. If a seal should fail because the motor is not submerged, the repairs are relatively simple, since the unit is easily accessible. The lower running speed of the PC pumps reduces wear and extends the periods between routine maintenance. Maintenance benefits are further enhanced by the unique flexible shaft design that simplifies the pump design by reducing the number of moving parts, thereby increasing reliability. This unique system eliminates wear movement between the drive end and the pumping element, making lubrication unnecessary. The pump manufacturer's confidence in this flexible shaft design was supported by a three-year warranty.

Although submersible centrifugal pumps were originally considered as a cheaper option, they were rejected because any cost benefits would have been short-term. Pump and motor maintenance would have been more difficult and time consuming, and the available seal types were limited.

Since being installed, these eight pumps have effectively handled the storm water runoff from the site.

Installation Case 7: A Sludge Application

Progressing cavity pumps are renowned for providing simple answers to difficult pumping problems. At Severn Trent's Priest Bridge Sewage Treatment Works in Brornsgrove, U.K., a progressing cavity pump has been operating efficiently on a duty where many other types of pumps would have failed (Fig. 63).

Surplus activated sludge is sent to gravity settlement tanks where, after the supernatant liquid has been drawn off, it is collected by tanker and transported to a digestion plant at Severn Trent's sewage treatment plant. In 1993, just prior to this installation, up to seven tankers were used each week to transport the surplus sludge.

As part of a general plant upgrade, a centrifuge sludge thickening system was installed to separate liquid from the

FIGURE 63
Installation example 7 (courtesy Monoflo Pumps Co.).

sludge more effectively, creating a thickened sludge with 6–8% dry solids content. This enabled much more efficient transportation—an average of only one tanker per week is now needed—cutting transport costs significantly. Severn Trent also stipulated to the pump contractor, Cullum Detuners, Ltd. of Derby, U.K., that the required fill rate was 45 liters per second to fill a 5,000-gallon tanker in just over eight minutes.

A major problem in pumping the sludge from the holding tank to the road tanker was the complex rheological behavior of the sludge, which exhibits Bingham fluid (plastic) characteristics. The extremely viscous nature of the sludge, together with a 7.2-meter static head and the presence of gas bubbles within the sludge, stretch the capabilities of many other types of pumps.

Cullum Detuners found the solution to be a vertically-mounted standard PC pump with a cast-iron body, a molded nitrile rubber stator, and a tool steel rotor. This single-stage PC pump is fitted with a large, 75 kW motor required to handle the high breakout torque created by the viscosity of the sludge.

A key feature that enhances the reliability of the pump in this arduous application is a special flexible shaft drive that provides a unique solution to the problem of connecting the rotating pump drive shaft to the eccentrically orbiting helical rotor. This eliminates the need for conventional universal joints. Because there are no wearing parts in the drive train, the maintenance costs associated with other forms of coupling are eliminated and service intervals are significantly lengthened. The simplicity of this flexible shaft drive and pump design means that the pumps can be easily dismantled and assembled when routine maintenance is required. The interference fit between the rotor and stator creates a positive seal that enables the pump to exhaust the gases produced by the active sludge during periods between pumping.

Since being commissioned in June 1993, this PC pump has proved its efficiency. It pumps the sludge at a rate of 45 liters per second, satisfying the criterion for loading a tanker in just over eight minutes. Its success has resulted in inquiries for similar applications elsewhere.

Installation Case 8: A Sludge Treatment Application

Progressing cavity pumps have increased the efficiency of sludge treatment at Northumbrian Water's Lartington Sludge Treatment Works in Cotherstone, U.K. (Fig. 64). Rain water

FIGURE 64
Installation example 8 (courtesy Monoflo Pumps Co.).

from the Grasholme and Hury reservoirs in the Tees Valley gravitates to Lartington Water Treatment Works (WTW), where it is treated and distributed throughout the Tees Valley, principally to Darlington and Stockton. Part of this treatment involves the removal of peat-based solids from the water. The resultant sludge is sent up to the Lartington Water Treatment Works at Cotherstone, U.K.

This sludge, which contains around 1–2% solids, is then pumped to centrifuges that extract most of the remaining water. The extracted water is sent back down to Lartington WTW and the dewatered sludge, which has a solids content of 15–25%, is then distributed onto the surrounding land.

The efficiency of the centrifuges depends upon the effectiveness of the pumps that deliver the sludge to them. Lobe pumps were installed originally, but these proved expensive to maintain due to the excessive wear caused by the abrasive nature of the sludge.

As we have discussed earlier, abrasion resistance and the ability to pump viscous fluids with ease are key benefits of PC pumps. Accordingly, the lobe pumps were replaced with single-stage PC pumps. These pumps do not require the critical clearances necessary between the lobes and casing of a lobe pump. In addition, the rolling action of the metal rotor within the rubber stator of a PC pump produces a pump well suited to handling particles in liquid suspensions. The fact that the contact area between the rotor and stator is constantly changing means that any particles trapped by the stator only remain so momentarily before being released.

Reliability is further increased by the flexible shaft drive (Flexishaft®), unique to this particular manufacturer. This links the drive shaft to the helical rotor, eliminating the need for universal joint designs. Because there are no universal joints, and therefore no wearing parts in the drive train, maintenance costs for these pumps are kept to a minimum. Periods between routine maintenance are therefore lengthened considerably. When servicing and maintenance is required, the reduced number of mechanical parts enables the pumps to be dismantled and assembled with ease. This pump manufacturer supports the reliability of its Flexishaft® with a three-year warranty.

Since the first of the four pumps so far installed were commissioned in 1994, these pumps have performed reliably for eight hours a day, five days a week, without the need for maintenance. The efficiency of the pumps has kept the four centrifuges working to their maximum capacity.

Installation Case 9: Reliability Is Key at a Food Processing Plant

Progressing cavity pumps are being used by Associated British Foods (ABF) to maintain the flow of a range of wheat starches,

FIGURE 65
Installation example 9 (courtesy Monoflo Pumps Co.).

syrups, glutens, and animal feeds at its plant in Corby, North-ants, U.K. (Fig. 65). ABF manufactures a range of products used in the brewing, baking, pharmaceutical, building products, agriculture, and paper industries. The company installed its first PC pump when the plant opened in 1982.

Locally grown wheat is milled into flour at Corby and then transferred to the main plant for processing into a wide range of products, varying in consistency from liquids to thick pastes and slurries. Equipment efficiency is crucial to maintaining production levels, as the plant operates 24 hours a day, 365 days a year, processing 3,000 tons of wheat each week. Reliability is thus a key requirement for pumps in this application.

ABF uses a combination of a standard inlet configuration design and an enlarged-throat hopper pump design version. The standard design with its unique drive system (Power-Drive®) is used to transfer water, white glucose syrup, liquid starch, and starch slurry between processes and from storage into tankers for delivery.

The PowerDrive® provides a single component link between the concentric motion of the bearing shaft and the eccentric motion of the helical rotor. The reduced number of moving parts in the drivetrain eliminates wear and makes lubrication unnecessary. The possibility of product contamination in the event of a joint failure is eliminated, a considerable advantage in the food industry.

The standard design is also used to encourage circulation within storage containers, because the glucose syrup layers out if adequate movement is not maintained. The hopper design pumps are ideal for pumping high-viscosity substances at ABF, where they ensure that gluten, used by the baking industry, and protein slurry, used in the production of animal feeds, are conveyed efficiently between processes. The enlarged-throat hopper design ensures that the thick, nonflowing gluten maintains a smooth, even flow.

Maintaining sanitary conditions during the process is essential at ABF. A comprehensive cleaning program is undertaken at regular intervals ranging from once a day to once every few weeks. The simplicity of these PC pump designs reduces the number of parts in the pumps, resulting in increased reliability, lengthening the period between routine maintenance, and making dismantling and reassembling easier for cleaning and maintenance purposes.

Installation Case 10: Sewage Treatment at an Airport

The world's largest progressive cavity pump was installed as part of a program to remove five million tons of accumulated sewage sludge from a site bordering Heathrow Airport in London, as quickly as possible (Fig. 66).

Decommissioning of Thames Water's Perry Oaks Sludge Treatment Works at Heathrow began in 1989, managed by land reclamation and recycling specialists Drinkwater Sabey CWD. Mogden Sewage Treatment Works delivered each day 17,600 U.S. gallons of anaerobically digested sewage sludge with a 2% dry solids content. This sludge was pumped into one of twelve lagoons at the site. After sedimentation occurs

FIGURE 66
Installation example 10 (courtesy Monoflo Pumps Co.).

and the excess water is decanted, the sludge has a dry solids content of 9–10%.

The pump at Perry Oaks empties the lagoons by pumping the thickened sludge at 200 tons per hour at a pressure of 87 psi. Operating 24 hours a day, the pump is stopped only for routine maintenance such as oil and diesel checks on the prime mover.

To accomplish this task, a PC pump was applied, which had a unique flexible shaft (Flexishaft®) design that links the bearing shaft to the helical rotor—a simple solution to the complex engineering problem of connecting the concentric motion of the pump drive shaft to the eccentrically orbiting rotor. To prevent stator damage, the sludge initially passes through a device called a "muncher" to grind any potentially damaging objects, such as stones and wood. Flow rate is important in this application: the pump's relatively slow speed, approximately 100 rpm, decreases wear and tear, providing a long working life and reduced maintenance costs.

The purpose of the pump is to move pump-thickened concentrate into the two largest lagoons (both holding over 1,760,000 U.S. gallons) on site. From these two lagoons, the sludge is transferred to tankers for distribution to local farms for use as a general purpose organic fertilizer.

Installation Case 11: Offshore Supply Vessels for Drilling Mud

Offshore supply vessels deliver drilling "mud" to drilling platforms in the Gulf of Mexico. Drilling mud is a mixture of clay, water, weighting material, and chemicals, used to flush cuttings from the drill bit to the surface. Traditional transfer of the mud from the supply vessels to the drilling rig has been by

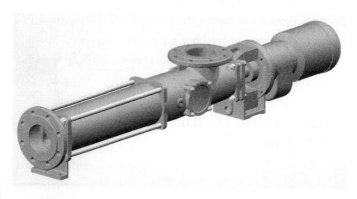

FIGURE 67
Installation example 11 (courtesy Colfax Pump Group).

use of centrifugal pumps (Fig. 67). Transfer flow rates are on the order of 300–500 gpm at discharge pressures of 125–225 psig. The mud is formulated to achieve a design-specific gravity. An SG of 1.7–2.3 would be common with a pumping viscosity of about 100 Cp and a solids content of 30%. Particle size of the solids would be in the 0.002–0.003 in. range.

The centrifugal pumps operate at 1,750 rpm at an efficiency of about 40%. A corresponding size two-stage PC pump would operate at less than 300 rpm with an efficiency of 75%. This results in a smaller, lower cost electric motor and less power consumption as well as relatively constant flow regardless of discharge pressure, low pulsation, metered flow, lower shear of the mud, and longer mean time between failures (MTBFs). The centrifugal pump must necessarily run at a relatively high speed to generate the required head, which runs counter to a long service life on the abrasive mud; the PC pump, like all rotary, positive displacement pumps, generates flow, independent of head (pressure). Shaft sealing for a PC mud pump is usually packing rings on a replaceable shaft sleeve. Most European supply vessels have used PC pumps for mud transfer for many years.

Installation Case 12: Main Feed Pump Plus Additives

An interesting oil additives application was a system that used a single motor to drive two pumps (Fig. 68). The main pump, driven by the motor and a gear box, pumps oil from the storage tank to the process. A certain amount of additive is re-

FIGURE 68
Installation example 12: The main oil pump module with the small additives pump, driven by a single motor, maintains a constant concentration of additives in pumped oil (courtesy Monoflo Pumps Co.).

quired, which must be maintained as a constant percentage of oil, while the overall demand changes from time to time.

The auxiliary small pump is belt-driven off the main shaft, and thus delivers a prescribed amount of additive to the discharge pipe of the main pipe. This resulted in a compact design requiring no auxiliary metering devices for the additive.

Installation Case 13: Vertical Orientation

The usual limitation of PC pumps is floor space, due to their length. However, the vertical distance is often a free dimension.

FIGURE 69
Installation example 13: A vertically oriented PC pump installation (courtesy Monoflo Pumps Co.).

In such instances, there are no reasons not to install the pump vertically, as shown in Fig. 69, as long as engineering determines that the rotor weight can be carried by the pump bearings. In most cases, this is usually not a problem, although it needs to be verified for each particular case.

Chapter 8

TROUBLESHOOTING

When properly selected and applied, PC pumps rarely are a cause of trouble and are rather forgiving in most cases. Even when a problem arises, it is usually a straightforward matter to identify the cause. For example, if there is no liquid delivered, the cause could be the same as for any other pump type—a clogged inlet line, wrong rotation, or air leaks in the inlet line.

If a PC pump stator is worn out, the slip may be excessive and eventually all flow is slipped back to suction. The wear can be caused by excessive discharge pressure (as a rule of thumb, 75 psi per stage is selected for pump sizing); dry-running; pipe strain and improper alignment; or too high a rotational speed. The issue of abrasion wear is controversial with regard to PC pumps because many users maintain that PC pumps' claim to fame is exactly that—their ability to handle abrasive substances (Fig. 70).

Therefore, if the failure rate becomes excessive for a particular application, users often expect warranty repair and a quick

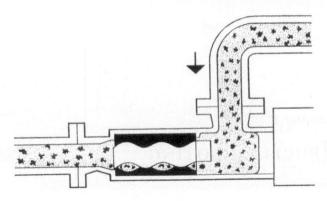

FIGURE 70
Handling of abrasive solids is a much-touted advantage of PC pumps.

solution to what they assume is pump problem. The reality is that, despite its excellent ability to handle abrasives, a PC pump can do this only if running at relatively low speed. The higher the concentration of solids, the slower the pump should operate. This is why it is important to specify the solids concentration, size, and hardness during the initial pump selection. If in doubt, it is always better to select a larger pump size; this may cost more initially but would significantly extend the stator elastomer life.

PC pumps usually run quiet. If a pump is noisy, it may be starved or might have air or gas leaks at the inlet pipe. Excessively high speed could also cause noise, as could improper mounting. Since the rotor eccentricity makes unbalance unavoidable, low rpm becomes especially critical for large pump sizes, such as shown in Fig. 71. Also, a heavy duty base plate and proper supporting structure, with anchoring at several stations along the pump length, are critical to keep vibrations down.

FIGURE 71
Large pumps require attention to the design of their base plates and supporting structures.

If the pump takes too much power, the fluid viscosity could be too high (i.e., higher than was assumed during the pump selection and sizing). Higher operating pressure would obviously also require more power because all rotary pumps produce the same flow per revolution of the shaft (i.e., power consumption at a given pump speed is directly related to differential pressure). Incompatible liquids may cause elastomer swelling, increased interference between the rotor/stator fit, and higher torque and power requirements.

In transfer applications, when distances are long and line friction losses become high, pumps must work extra hard to overcome excessive differential pressures. As we learned in Chapter 3, more stages are required, since each stage is designed to 75 psi (or less if abrasives are present in significant concentration). A PC pump can become too long, resulting in manufacturing problems. But even not counting manufacturing problems, a very long rotor becomes too flexible and tends to sag in the middle, resulting in excessive interference with the elastomer lining, overheating, and failure ("chunk-out"). It is sometimes possible to use several pumps, each pumping from into an intermediate storage open vessel from which the next pump takes suction, as shown in Figs. 72 and 73. This ensures that inlet pressure for all pumps is the same and not increasingly high.

For applications where the pumpage is appreciably corrosive, the stator lining can be made from an elastomer with better corrosive resistance than standard rubber. Teflon® lining is sometimes employed for this reason. The elastomer layer prevents the chemicals from reaching the stator metal, which is usually a plain cast iron. However, if the condition of the stator ends deteriorates and they are simply cut off (which is a standard procedure), pumpage will penetrate through the interface at the ends and eventually work its way in, dissolving the bond between the elastomeric lining and the stator tube, causing eventual bond failure. A proper design should have the elastomer ends extend completely beyond the tube and flare out, creating a positive sealing layer to prevent chemicals

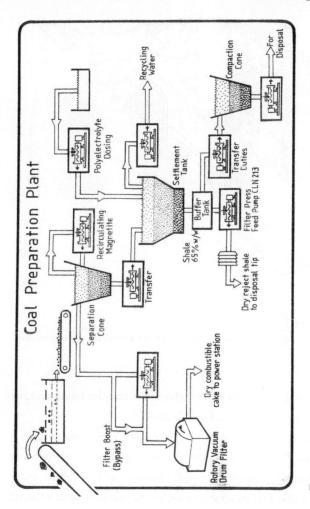

FIGURE 72

A coal transport application where several smaller (shorter) pumps have been successfully used.

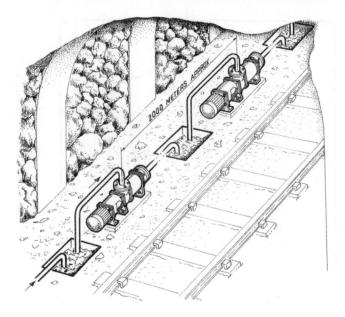

FIGURE 73
Several pumps in a row move coal slurry via a sequence of intermediate ponds.

from penetrating into the interface between the lining and the stator tube, as shown in Figs. 74 and 75.

When a new pump is considered as a replacement unit, it might be desirable to first try it out on a temporary mounting skid, thus avoiding costly piping changes and installation foundation work. Figure 76 shows such a mobile unit, which can be mounted by maintenance personnel or supplied complete by the manufacturer.

This could be especially attractive for small pumps where unbalance forces are smaller and the need for a permanent

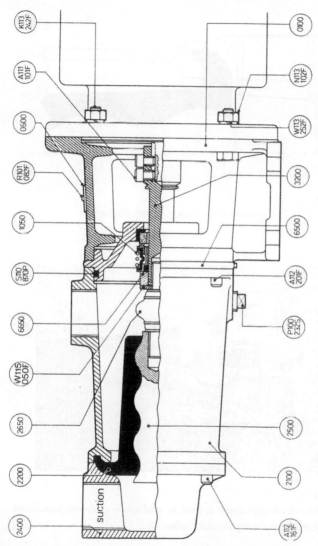

FIGURE 74 *An elastomer must flare at the ends to completely cover the interface between the liner and the stator tube to prevent pumpage from creeping into the interface, causing bond failure.*

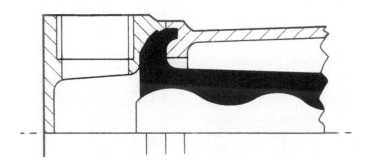

FIGURE 75
Detail of the lining at the stator ends, properly flaring.

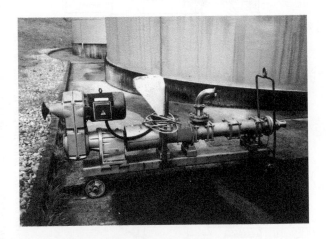

FIGURE 76
A mobile trial unit is simple to hook up and avoids expensive foundation and piping work during the test evaluation.

foundation is not as critical, especially for a temporary trial. An arrangement for a 30- or 60-day trial period can be part of the warranty, and the user would return the unit if it failed to meet the specified operating conditions and guaranteed life.

Even though PC pumps have excellent NPSHr character- istics and can self-prime very well, they may gradually lose these benefits after wear in the field. Wear increases clear- ances between the rotor and stator and allows air to circulate back to suction through these clearances, thus preventing suc- tion line evacuation (priming). This results in dry-running; if this continues for more than 30–60 seconds without the pump getting primed, it can easily fail. Figure 77 shows a pump that takes suction with a lift condition.

After each restart, the pump must self-prime or be primed manually. A better system design would connect suction pip- ing below the liquid level to ensure flooded suction conditions at all times, similar to an installation shown in Fig. 78.

There are cases, however, where top tank suction is required and a lift application is a must. This is typically the case for dan- gerous or expensive liquids, where a failure (breakage) of a suc- tion line would result in complete loss of the tank contents on the ground if the pump suction is flooded. Railroad cars, for example, are typically unloaded via the top-suction method, and self-priming pumps are used. Positive displacement pumps are inherently self-priming, while centrifugal pumps are not (with the exception of specialized self-priming designs).

Certain substances with extreme viscosities (e.g., clay) are very difficult to get into pumps, and suction design is critical.

FIGURE 77
Pump suction with a "lift" condition, needing priming or self-priming.

FIGURE 78
Pump suction is always flooded, ensuring no need for priming.

PC pumps are sometimes installed directly underneath tanks and equipped with hoppers to make sure the pumpage falls directly into the pump suction opening by gravity (Fig. 79).

Even for less difficult substances where the pumps are not directly under the tanks, the use of hoppers is a good idea because they channel the pumpage to the inlet, as shown in Figs. 80 and 81.

Even with hoppers, certain substances have trouble getting from the bottom of the hopper to the pump inlet to begin filling the rotor-to-stator cavity. Consider an example of an application at a sewage treatment plant (Figs. 82 and 83).

The incoming sewage undergoes a series of treatments and the separated raw sludge goes to the incinerators (Fig. 84), while the thinner substance is pumped on for further treatment.

FIGURE 79
Pumps equipped with hoppers mounted directly underneath tanks receive a process substance that falls down by gravity.

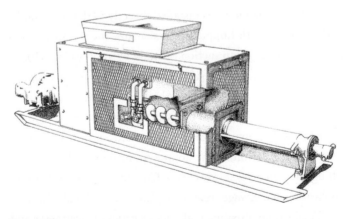

FIGURE 80
A hopper helps guide the process pumpage to the pump inlet.

FIGURE 81
Olives are channeled into a pump hopper in an olive oil application.

FIGURE 82
A sewage treatment plant.

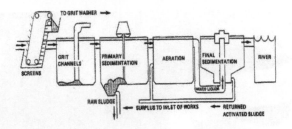

FIGURE 83
Sewage works flow chart.

Since this plant was experiencing problems with thickened sludge bridging at the pump inlets, new designs for the plant module expansion (Fig. 85) called for pump inlet augers, which are featured in Figs. 86 and 87. These enhancements solved the problem.

When troubleshooting a pump, it is critical to know how it performed originally at the factory. Without such a test, one can never be sure if the problem is pump-related or system-related. It may take a long time to troubleshoot the system only to find out, in the end, that pump rotor-to-stator clearances were applied incorrectly. While such mistakes may be relatively rare due to quality assurance procedures implemented at most pump manufacturing facilities, the best assurance of successful pump operation is still, as a minimum, a factory test. For critical applications such as drilling rigs, a failure of the mudmotor during drilling is very costly and time-consuming.

Production logistics of mudmotors are different from PC pumps. While PC pumps are manufactured and shipped complete and ready to install, and are produced by a single pump

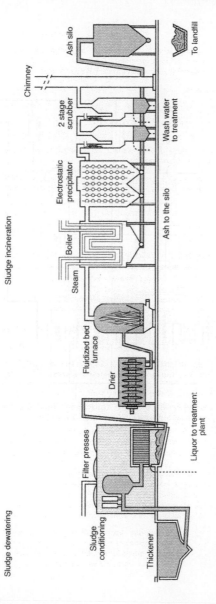

FIGURE 84 *A typical incinerator schematic.*

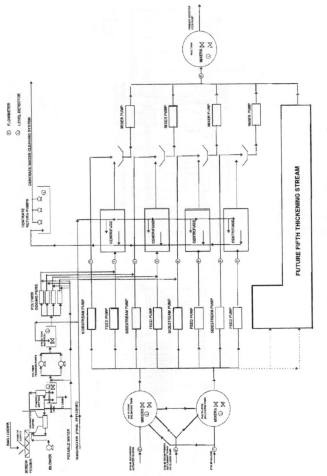

FIGURE 85   *A typical sewage treatment unit, undergoing design expansion.*

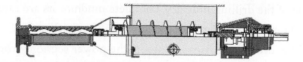

FIGURE 86
An inlet auger helps move the pumpage to the pump inlet cavity.

FIGURE 87
A pump equipped with an inlet hopper and auger for pumpages that are tough to move to the inlet.

manufacturer, mudmotors are usually made by several producers. The rotor and stator are made by one manufacturer (usually the same one who also makes complete pumps), but the rest of the power string (joints, supporting bearings, seals, and drill bits) are assembled by other producers who are usually located closer to the drilling production areas. In the present mode of the drilling industry, complete mudmotors are rarely tested until they are actually installed on the rig (Fig. 88).

However, the critical element (the power section—rotor plus stator) needs to be, and usually is tested by either the original manufacturer or the integrator, or both. Figure 89 shows a

FIGURE 88
A mudmotor being worked at the drilling rig.

FIGURE 89

A mudmotor power section (rotor plus stator) is tested for torque versus pressure characteristics at several speeds.

test of the mudmotor power section to obtain torque versus pressure characteristics at several speeds.

Downhole pumps (Fig. 90) are somewhat middle-of-the-road in terms of their manufacturing logistics.

Downhole pumps (DHPs) are used primarily in the oil industry for oil recovery. Their designs, from a technical point of view, are much simpler than either PC pumps or mudmotors. In terms of production volume, DHPs are a relatively small segment among all PC machines, probably not significantly more than 5% of all units.

FIGURE 90
A downhole pump cross section.

Downhole pump test characteristics, obtained during performance testing, are available at the manufacturer's Test or Quality Assurance (QA) department. It is also advisable to obtain either a standard catalogue-published performance curve for existing designs, or engineering estimate curves for a new design, as well as as-tested curves for the completed design. It is a good idea to visit the pump or mudmotor manufacturer and examine their QA-related certificates (see the example of a 3-A certificate to supply equipment for milk and milk products, Fig. 91).

Such certification is not just a formality; it can actually directly impact our lives (Fig. 92).

Initially issued Oct 22, 1991
U.S. Representative:
MonoFlo, Dresser Pump Div.
821 Live Oak Drive
Chesapeake, VA 23320-2601

Authorization No. 654

This Is To Certify That

Mono Pumps, LTD

Martin Street, Audenshaw, Manchester M34 5DQ ENGLAND

Is hereby authorized to continue to apply the 3-A symbol to the models of equipment, conforming to the 3-A Sanitary Standards for Centrifugal and Positive Rotary Pumps for Milk and

Milk Products

(02-08A1)
_____ set forth below:

Model Designations "S" Range of Hygienic Pumps

_____ for the twelve months ending October 21, 1997

Date of Issuance: Oct 22, 1996   *Earl O. Wright*, Secretary
3-A SANITARY STANDARDS SYMBOL ADMINISTRATIVE COUNCIL
★ ★ ★ ★ ★

The issuance of this authorization for the use of the 3-A symbol is based upon the voluntary certification, by the applicant for it, that the equipment listed above complies fully with the 3-A Sanitary Standards designated. Legal responsibility for compliance is solely that of the holder of this Certificate of Authorization, and the 3-A Sanitary Standards Symbol Administrative Council does not warrant that the holder of an authorization at all times complies with the provisions of the said 3-A Sanitary Standards. This in no way affects the responsibility of the 3-A Sanitary Standards Symbol Administrative Council to take appropriate action in cases in which evidence of non-compliance has been established.

FIGURE 91
An example of a 3-A certificate.

FIGURE 92
These products might have come to our homes with the use of 3-A certified PC pumps.

While visiting a pump manufacturer, it is good to inspect the production facility, from the rotor and core milling (Fig. 93), to polishing and storage (Fig. 94), to stator elastomer injection operations (Fig. 95).

It is also critical to ensure that the Quality Assurance department maintains a log of rotor and stator final measurements, with good traceability to the shipped units (Figs. 96 and 97).

FIGURE 93
A rotor milling operation.

Overall, PC pumps have very broad applications and new applications are constantly being evaluated and applied. Their versatility and robustness are rapidly gaining reputation as problem-solvers, and their designs are becoming better understood and enhanced. However, the complexity of the geometry of their hydraulic section (the rotor inside the stator, creating the complex cavity) is still the main stumbling block for users, designers, and contractors. This complexity of the geometry does not, however, necessarily translate into complexity in

FIGURE 94
Storage of cores used for elastomer-to-tube-molding.

maintenance. A well-designed progressing cavity pump should not present difficulties in assembly and maintenance (Fig. 98).

Pressing the rotor into the stator is easier if it is lubricated with soapy water or oil. For tight interference fits, PC pumps sometimes experience startup difficulties, and oversized motors are recommended in these cases. If possible (especially with more viscous fluids), it is better to have a loose fit (clearance) between the rotor and stator; this will significantly improve wear and life, and will make starting easier.

FIGURE 95
A stator elastomer injection operation.

Technical papers on these issues are just beginning to emerge at conferences and in journals. At the present time, very little technical information has been published pertaining to design, and most of the knowledge is confined to a few manufacturers' engineering departments. Even this information is often in empirical form, without too much dependency on the first principles. Since the original works of Rene Moineau, most research efforts have focused on application aspects rather than basic design optimization. There have been several exceptions, however, and new design features

FIGURE 96
A rotor on a coordinate measuring machine.

(e.g., flexible shafts) and advanced engineering tools have begun to surface (Fig. 99). Such techniques aid in troubleshooting to identify hot spots and excessive tendency to wear due to tight clearance, thermal expansion effects, and so forth.

Therefore, the time has arrived in the pump industry to summarize the status and place of PC pumps among the other pump types, along with the need for a comprehensive technical book to tie together nomenclatures to explain (in simple terms) the principles of operation and to provide basic refer-

Figure 97
Stator dimensions are critical, as the elastomer molding process is more difficult to establish and control as compared to metal rotors.

ences for application-related inquiries. The following sections of this book, therefore, establish a basis for pump design philosophy and set up a structure and mechanism for future enhancements and developments. We are continuing to gather and monitor feedback from the pump design, user, and application communities so that future editions of this book can present new ideas, input, criticism, and suggestions to further explain, promote, and apply these types of pumps.

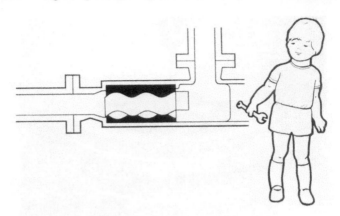

FIGURE 98
A well-designed pump should be easy to work on during maintenance.

FIGURE 99
CAD-generated model of a PC pump.

Troubleshooting PC pumps in systems is a skill that takes time to develop. The guidelines in this book were established to reduce that time while providing a framework for determining the cause of a problem.

In almost any pumping system, the pump is the most vulnerable component. Therefore, regardless of what may actually be wrong, the symptoms frequently suggest what seems to be a pump problem or they cause damage to the pump that is then viewed as the culprit. Most reputable pump manufacturers do not produce very many defective pumps. The problem is usually caused by a component malfunction, inadequate control of the pumpage, or a change in operating requirements to which the system or pump cannot adapt.

Identifying which of these issues is the root cause is not always easy, but an open mind certainly helps. There are mechanical and hydraulic aspects that may relate to a cause of a problem. For example, loss of flow from the pump could be caused by a broken driver shaft (a mechanical cause) or by cavitation caused by inlet pressure losses due to viscous liquid (an hydraulic cause).

Here, we make no provisions for special industry or regulatory practices. Some pumping systems handle toxic, corrosive, flammable, or other dangerous liquids and thus need special precautions that go beyond the scope of this book. Some rotary pump systems are life-supporting or operate in critical or sanitary services where malfunctions can have dire consequences. Be sure to understand the repercussions of troubleshooting activities on the system as well as the personnel involved.

Before You Begin Troubleshooting

Before beginning the troubleshooting process, be sure pressure gages are available for pump inlet and discharge. A temperature gage at the pump inlet can also provide valuable information, depending on the application or problem. There should be some method of verifying pump speed via either a hand-held tachometer or strobe tachometer if the system is not already so equipped.

Also, be sure that the drivers are absolutely locked out and tagged out before removing guards or conducting any system inspections. If the system includes accumulators, pulsation dampeners, or other spring or compressed gas energy storage devices, be certain they are fully discharged of liquid before doing any system work. Observe all company and industry safety regulations and guidelines to minimize any inherent hazards to personnel or the system.

In this review, we will seek to uncover problems in existing systems that had been operating satisfactorily. This approach is somewhat different from a first-time startup of a new system where issues such as reverse-installed valves, missing components, wiring errors, extensive fabrication debris, pipe strain, etc. are the norm.

Information Gathering

Make note of anything that has changed since operation was last satisfactory, regardless of whether it seems to be unrelated to the problem. Was the system undergoing routine maintenance? Were any new or repaired components changed out? When was the pump last serviced, and what was that service? What were the appearance and condition of the pump's internal parts? From where were replacement parts obtained? For recirculating systems (such as for lubrication), was new or additional liquid added? For one-time-through systems (such as additive injection), did the additive supplier, additive grade, or additive temperature change? How long did the pump operate before the problem developed? Describe the problem in as simple and straightforward a manner as you can.

Write down the normal expected pump operating conditions as follows:

1. Inlet pressure

2. Outlet pressure

3. Flow rate

4. Pump speed

5. Liquid

6. Minimum and maximum liquid temperature

7. Minimum and maximum liquid viscosity

8. Continuous or intermittent duty cycle

Note the presence or absence of a change in pump noise or vibration since it last operated satisfactorily. A change in either or both of these characteristics points to a number of very specific things to check as possible causes.

Flow Loss or Low Flow

Be sure the pump's direction of rotation is correct; this is an obvious but frequently overlooked problem. Be sure the operating speed is correct (this is especially true for drivers that can operate at more than one speed, such as variable speed drives, turbines, engines, etc.). Flow loss is normally accompanied by a reduction in system pressure. Either the pump is delivering less flow or the system is bypassing it, such as through a defective or worn relief valve, bypass valve, or pressure control valve. The pump could be worn and internally bypassing (slipping) flow so that less flow reaches the system. In that event, pump repair will be necessary. A partial inspection of the pump internal components will usually reveal wear conditions. If the operating viscosity of the liquid has been reduced (a new liquid or a higher operating temperature), a rotary pump's rated flow will be slightly reduced due to some of the flow "slipping" back to suction (i.e., recirculating), and this will be even more so for higher pressure operations.

Loss of Suction

Loss of suction can be minor, causing little damage, or could be a major cause of catastrophic damage. Loss of suction means that liquid flow is not reaching the pumping elements or not reaching it at a sufficiently high pressure to keep the pumpage in a liquid state as the pumping elements capture a volume of liquid. Loss of suction can be caused by the pump being unable to prime, by cavitation, or a gas content problem.

PC pumps are self-priming. That means that, within limits, they are capable of evacuating (pumping) a modest amount of air from the inlet (suction) system into the discharge (outlet) system. However, rotary pumps are frequently not very good air compressors and the pump discharge should be temporarily vented to allow the inlet side air to escape from the discharge side of the pump at low pressure. If possible, filling the inlet system with liquid, or at least filling the pump (wetting the pumping elements), will make a major improvement in the pump's priming capability.

Cavitation is insufficient system inlet pressure to the pump; this pressure is to prevent the liquid from partially vaporizing. This can be caused by an inlet system restriction, excessive liquid viscosity, or excessive pump speed. Inlet restrictions can include dirty or clogged inlet strainers, debris floating in the liquid supply that covers the inlet piping intake, or rags or port blanking flanges that have gotten into the system or were not removed, especially after maintenance. If the liquid is cooler than the design temperature, its viscosity (thickness) may be too high, causing excessive friction (pressure loss) in the inlet piping system. In the latter case, it may be necessary to increase the pumping temperature of the liquid. If the liquid being pumped has changed, a change in its viscosity-temperature characteristics or its vapor pressure-temperature characteristics should be examined. Cavitation is caused by operating the system inlet pressure-temperature in a combination such that the liquid's vapor pressure (a pressure at which the liquid converts to a gas at pumping temperature) is reached. The liquid begins to partially vaporize and the pump is unable to handle this compressible vapor-liquid mixture.

Cavitation is frequently accompanied by noise, vibration, and significant increase in discharge pressure pulsation. Modest cavitation will cause pitting to appear on pumping elements, not unlike that found near the root of marine propeller blades.

Entrained gas (as compared to liquid turned to vapor) in the inlet flow has the same impact on pump operation and the same symptoms as cavitation with regard to loss of flow, vibrations, etc. But gas does not cause pitting damage, as compared to vapor. It can be caused by vortexing (whirlpooling) of the liquid in its supply source, which drags air into the liquid. If a pump operates at an inlet pressure below local atmospheric, it is quite possible that air is being drawn into the inlet piping through a loose piping or pump casing joint, a leaky suction valve stem, or a defective, cut, folded or otherwise damaged inlet system joint gasket. In recirculating systems (such as a lubrication system where the liquid pumped is continuously returned to a supply source or tank), if the tank and return lines are not adequately designed, located, and sized, air is easily entrained in the liquid and immediately picked up by the pump inlet system. Be sure the liquid level at its source is at or above minimum operating levels. Lines returning flow to a supply tank should terminate below the minimum liquid level. Internal tank baffles may be necessary to provide full tank volume retention time so any entrained air can more readily dissipate.

Low Discharge Pressure

Pump discharge pressure is caused by the system's resistance to the flow provided by the pump. If it is low, then either the pump is not providing the expected flow or the system is not offering the expected resistance to that flow. It is possible that flow is being restricted into the pump (cavitation or suction starvation). This is usually accompanied by noise and vibration, the pump is not producing its rated flow (the pump is worn or damaged or its drive speed is too low), or the pump flow is bypassing rather than being delivered into the system as intended (possibly caused by an open, improperly set, damaged, or worn discharge system valve). If the pump is relatively new and not in an abrasive application, it is most probable that the discharge flow is bypassing. The most likely paths for such unwanted bypass are: the system pressure relief valve; a leaking bypass pressure regulator; an inadvertently open bypass valve; or any of these valves having worn valve seats, incompletely closed stems, incorrect signal controls, broken springs, etc.

Many pumps can be quickly, if incompletely, inspected in place without disturbing piping or the pump alignment. If the pump does not smoothly turn over by hand or with a little leverage assistance, the pump may indeed be the problem. To visually inspect the pump internals, the pump stator will need to be removed to expose its inside condition as well as the condition of the rotor. One should be able to readily see enough wear to cause a pressure reduction (flow loss).

It is sometimes difficult to determine if a valve is bypassing when it should not be. It is probably best to remove the valve, partially disassemble it, and examine the mating valve seat surfaces or seat seals for wear or damage. Check any spring to be sure it is not broken. Work the valve mechanism manually, if possible, to detect any binding or galling.

If the problem has still not been identified, be sure the pump driver speed is being achieved and that the actual pump shaft is rotating at its correct speed. These are especially true of a new system startup.

Excessive Noise or Vibration

As already discussed, excessive noise and/or vibration can be a symptom of cavitation, suction starvation, or excessive gas in the liquid. This is especially true if the discharge pressure is fluctuating or pulsating. Mechanical causes of noise and vibration include shaft misalignment, loose couplings, loose pump and/or driver mounting hardware, worn or damaged driver or pump bearings, or valve noise that seems to be coming from the pump. Valves, especially on the discharge side of the pump, can sometimes go into a hydraulic vibration mode caused by operating pressure, flow rate, and the valve design. Resetting or changing an internal valve component is usually sufficient to solve the problem—consult the valve supplier. If the drive system includes gearing, belts, or chain drives, sheave and sprocket alignments become very important and should be checked.

Excessive Power Usage

Excessive power consumption can be caused by either mechanical or hydraulic problems. For PC pumps, the pump power requirements are directly proportional to pressure and speed. If either have increased, the required input power will also increase. Power required will also increase if the fluid viscosity has increased. This can happen if the liquid has been changed or the liquid operating temperature has been reduced. Some liquids are also shear sensitive (grease, for example) and can become more or less viscous with shear (pumping action); they might also undergo permanent viscosity change from shear over time.

Mechanical causes include bearing wear, rubbing pumping elements that lead to a pump failure, bad shaft alignments, and poor pulley alignments for belt-drive arrangements.

Rapid Pump Wear

Rapid pump wear is caused either by abrasives present in the liquid or operation at a condition for which the pump is not suitable, such as excessively low viscosity or excessively high pressure or high temperature. If abrasives are a normal condition of the pumping application, such as slurry pumping, then pump wear will be a fact of life. In such cases, the best that can be done will include pump and drive speed selection that provides the best economic value over the pump's lifecycle. While slower operation in abrasive service requires bigger displacement and more expensive pumps, it often pays back far beyond the initial purchase cost differential. Wear due to abrasives in the liquid is a function of speed raised to a power, usually between two and three. Obviously, if abrasive foreign material is not supposed to be present, strainers or filters should be employed wherever possible and practical.

Rapid wear is sometimes not wear in the sense of a pump being nondurable, but is really a catastrophic pump failure that occurred very quickly. Looking at the pump's internal parts alone frequently cannot provide much help in setting a search direction. This is why it is important to know what was occurring in the time period immediately preceding detection of a problem.

PC Components in Service

Figures 100–118 show the effects certain operating conditions can have on PC pump components.

PC pump rotors are normally interference fitted to their stators. The pump relies upon the presence of liquid to provide a lower-friction sliding contact and to remove the fric-

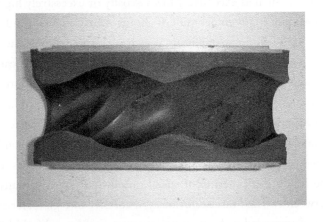

FIGURE 100
An axially cut single-stage stator that has been run dry. The right end is the suction end. Clearly, material has been torn from the ID surface.

tional heat from rubbing. In the absence of liquid, the rotor runs hard on the stator, causing local overheating and destruction of the elastomer. Stators can be equipped with thermal sensors (temperature switches) that can detect an increase in heat and sound an alarm and/or shut down the drive.

Strainers are recommended to keep foreign objects out of the pump or to at least limit the size that can get into the pump so as to minimize the chance of doing damage.

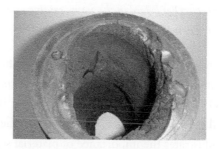

FIGURE 101
The discharge end of a severely dry-run stator.

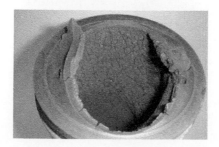

FIGURE 102
The suction end of the stator from Fig. 101.

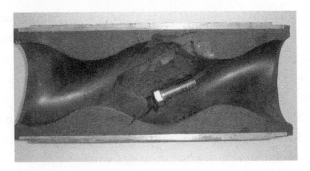

FIGURE 103
A large foreign object has lodged in the pump, causing extensive damage to the pump stator.

This is indicative of a chemical incompatibility between the elastomer and the pumped liquid. As the elastomer swells, the stator ID closes down on the rotor, increasing friction and heat that will cause further deterioration of both the stator and rotor. This illustrates the importance of material selection for pumping components.

FIGURE 104
The elastomer in this stator has swollen and now bulges from the end of the stator.

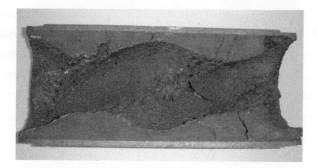

FIGURE 105
A section from a single-stage stator that has been exposed to chemical attack and mechanical abrading of the elastomer due to swelling.

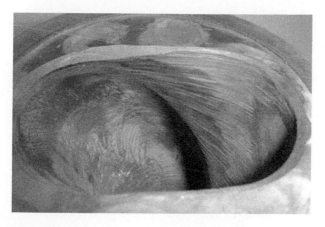

FIGURE 106
Typical long-term wear in a stator from fine abrasive content in the liquid.

The interference fit between the rotor and stator will gradually become a clearance fit. The pump flow rate will gradually decrease until it is no longer sufficient for the intended purpose. At that point, the stator will need replacing. The rotor should be examined and may be suitable for one more use with a new stator.

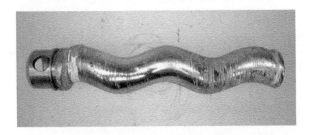

FIGURE 107
A well-worn rotor, no longer very usable.

FIGURE 108
A close-up of the discharge end of the rotor in Fig. 107. The end of the rotor is outside the stator and subject to very little wear. As can be seen from the size change in the OD of the rotor, the wear shown is extensive.

FIGURE 109
Similar to Figs. 107 and 108, but this rotor has been worn with much finer abrasive particles in the liquid. There is no heavy scoring but the diameter change at the discharge (right) end is apparent.

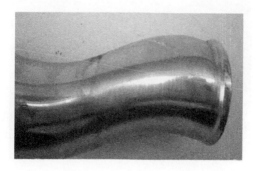

FIGURE 110
A close-up of the discharge end of the rotor in Fig. 109. The diameter change is very visible.

FIGURE III
Both a mechanical and chemical attack of the chrome surface of a rotor. An alternate rotor material may reduce or eliminate this problem.

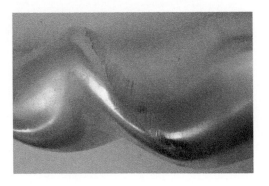

FIGURE II2
Some of the chrome plating has worn off this rotor where it is in contact with the stator.

A rotor in the condition shown in Fig. 112 can have the chrome stripped and reapplied to extend its useful life.

One end connects to the pump drive shaft while the other connects to the pumping rotor. The large holes on either end are for wear bushings and the drive pins that allow the eccentric motion of the rotor within the stator. As designed and manufactured, the holes are parallel. Figure 113 shows that the connecting rod has twisted nearly 90 degrees from a condition that jammed the pump. After the jamming cause is removed, this coupling rod will need to be replaced.

FIGURE 113
A connecting rod, sometimes called a coupling rod.

FIGURE 114
A connecting rod with hardened bushings installed. The left side is the drive side; the right side is the rotor side. The drive pin has worn through the bushing and into the connecting rod material. The connecting rod assembly will need replacement.

FIGURE 115
The hardened drive pin from the rotor in Fig. 114. This is from a large pump that had accumulated 22,000 hours of operation.

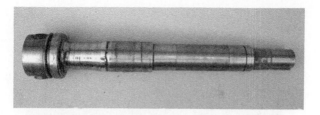

FIGURE 116
A drive shaft from a packed pump that has been destroyed from re-peated repacking without removal of old packing. Continuously tighten-ing the packing gland drove the packing rings into very hard contact with the shaft.

FIGURE 117
A close-up of the packing diameter of the pump shaft. Severe necking down of the shaft diameter could have resulted in a torsional shaft failure.

FIGURE 118
Normal wear of packing rings on a shaft sleeve mounted to the pump shaft. The shaft sleeve, while costing a little more, prevents wear from the packing rings on the pump shaft, a more expensive part.

There is a very strong tendency to place immediate blame on the pump. This usually causes other areas of the system to be overlooked and the real problem remains obscure. An open and inquisitive approach will usually determine the true root cause.

Chapter 9

PROGRESSING CAVITY PUMP SELECTION AND SIZING

If one reviews manufacturing statistics for most machinery, including pumps, it quickly becomes apparent that many more small machines are made than large ones. Smaller machines are made in larger quantities that justify investment in highly cost-effective manufacturing methods. Larger machines frequently cannot justify such investments and are avoided for this reason, if at all possible. The sheer volume of material needed for large machines also tends to drive their costs up. On the other hand, large machines tend to be more efficient than small ones.

So what has this to do with sizing a PC pump? Economics and competitive pressures will drive the selection toward the smallest possible pump size operating at the highest practical speed. There is nothing inherently wrong with this practice, provided it takes into account all the user system requirements, resulting in satisfactory, usually long-term, operation.

It is recommended that pump sizing be done by trained, experienced PC pump personnel. Computer programs are usually used only for preliminary selection, and PC experience determines the final choice among candidate pump sizes.

Pump selection and sizing should follow these steps:

- Specify requirements

- Select pump size, speed, geometry, and number of stages

- Select materials (rotor, stator, suction and discharge casings, connecting rod, covers, O-rings, gaskets, etc.), shaft sealing options (packing with or without flush, single, double, or tandem mechanical seals with auxiliary flush or barrier fluid circulation)

- Select type of speed reduction (gear, pulley, variable speed devices, etc.)

- Select driver (electric motor, combustion engine, turbine, hydraulic, or pneumatic drive, etc.)

The principal limiting factors restricting speed (which dictates size and cost) in PC pump applications are:

- System net inlet pressure available

- Rotor-to-stator rubbing speed

- Nature and volume of solids content in the pumped liquid

These limitations may vary with the manufacturer, design features, as well as the materials. In the presence of abrasive content, pump life is inversely proportional to a value between the square and the cube of the pump speed. Lower speeds will contribute to longer life but, for the same flow rate, the slower pump will be larger and almost certainly more expensive on a first-cost basis.

The American Petroleum Institute (API) publishes a standard, API-676, for rotary positive displacement pumps. It is a good source for pump specifications but, perhaps more important, it contains a data sheet to be completed by the user or buyer for inquiry purposes and the seller or manufacturer for quotation purposes. It collects the most relevant data needed to specify user requirements, the nature of the liquid to be pumped, and the conditions of pump operation. It is a good overview that covers all usual pumping needs, but it does not collect information about solids content. Therefore, add at least:

- Weight % solids
- Maximum particle size
- Maximum fiber length
- Description of solids, especially as it relates to abrasion

In classic pump selection, the system net positive inlet pressure available (NPIPa) must at least equal, and preferably exceed, the pump net positive inlet pressure required (NPIPr). If the reader is familiar with centrifugal pumps, the comparable terms are NPSHa (net positive suction head available) and NPSHr (net positive suction head required). For PC pumps that frequently deal with a nonhomogeneous fluid (i.e., having solids and/or gas or froth) delivered to the pump in clumps or batches rather than continuously, the NPIP concept becomes slightly less clear. Empirical data gathered by pump manufacturers provides guidance on maximum pump speed relative to keeping the inlet pump cavity reasonably filled before it closes off to inlet flow. Many PC pumps can be provided with an auger screw welded to the connecting rod (coupling rod) that force-feeds inlet fluid into the pumping elements. Independently driven auger screws can also be provided on hopper-style pumps that will force thick, high-viscosity solids material into the pumping screw.

Rotor-to-stator rubbing speed is the maximum surface speed of the rotor outside diameter as it rotates and oscillates within the stator. Since the rotor and stator normally have an interference fit at operating temperature, rubbing speed should be limited to about 16 ft/sec on clean, lower-viscosity liquids (no solids). This means that larger displacement pumps will need to be operated at lower speeds than small displacement pumps, a fact that applies to most all-rotary positive displacement pumps.

Example 1: Pump Sizing

Fluid: Secondary waste water sludge
Flow Rate: 100 gpm
Differential Pressure: 50 psid
Maximum Particle Size: 0.25 in.

First, categorize the degree of abrasiveness from Table 1. In our example, it is "medium." If in doubt, select the next-worst classification.

Table 1

Abrasives	Viscosity	Max. Rubbing Speed (ft/sec)	Max.* rpm	Fluid Examples
Heavy	Very high	<2	450	Filter press cake, tile grout, primary wastewater sludge
Medium	High	2 to 4	925	Secondary waste water sludge, wood putty, dry wall mud
Light	Medium	4 to 8	1,800	Baking batters, ground fish, cookie filling
None	Low	8 to 16	3,000	Polymers, fats, oils

*Abrasive or viscous speed limit (may be lower than mechanical speed limit).

From Table 2, the smallest pump that can handle the specified particle size is "K." Neglecting slip, the "K" size would need to operate at 100 gpm / 25.099 gal per 100 rpm × 100, or 398 rpm. At 398 rpm, the rubbing speed is 398 rpm × 1.95 ft/sec / 100 rpm / 100, or 7.76 ft/sec (Table 3). This rubbing speed far exceeds the recommended maximum range of 2–4 ft/sec for medium abrasive fluids (see Table 1). Examining the next several larger sizes as follows reveals the optimum selection as size "N," the smallest pump that operates nearly in the maximum recommended rubbing speed range and more than able to handle the maximum specified particle size.

TABLE 2

Size	Gal per 100 rpm**	Displacement (max. * rpm)	Displacement (max. * gpm)	Rubbing Speed (ft/sec) per 100 rpm	Maximum Particle Size (inch)	Maximum Fiber Length (inch)
A	0.053	3,000	1.6	0.22	0.03	1.0
B	0.099	3,000	3.0	0.30	0.04	1.2
C	0.198	3,000	5.9	0.42	0.04	1.4
D	0.396	3,000	11.9	0.49	0.06	1.4
E	0.793	2,500	19.8	0.64	0.08	1.4
F	1.651	2,000	33.0	0.79	0.10	1.7
G	3.303	1,600	52.8	0.98	0.12	1.7
H	6.605	1,200	79.3	1.33	0.15	1.9
J	13.210	1,000	132.1	1.57	0.20	2.4
K	25.099	800	200.8	1.95	0.27	3.1
L	36.328	700	254.3	2.21	0.27	3.1
M	49.538	650	322.0	2.46	0.37	3.9
N	66.050	575	379.8	2.66	0.37	3.9
P	95.773	500	478.9	3.15	0.55	5.1
R	178.336	425	757.9	3.84	0.79	8.3
S	330.251	330	1089.8	4.59	0.98	9.8
T	627.477	275	1725.6	5.71	1.18	9.8

* Mechanical speed limit.

** Theoretical (neglecting slip).

TABLE 3

Size	RPM per 100 gpm	Rubbing Speed (ft/sec)
K	398	7.76
L	275	6.08
M	202	4.97
N	151	4.02
P	104	3.28

TABLE 4

	Maximum PSID/Stage		
	Unequal Wall		Equal Wall
Abrasives	*1:2 Geometry*	*2:3 Geometry*	*1:2 Geometry*
Heavy	15	22	30
Medium	35	52	70
Light	60	90	120
None	87	130	175

A conventional stator is manufactured using common cylindrical pipe as the outer structure. It is strong and relatively inexpensive. However, when the elastomeric liner is injected between the pipe inside diameter and the removable core in the molding process, the wall thickness of the elastomer varies along the length of the stator from the minimum design thickness to perhaps twice that thickness. This type of stator is known as an unequal wall stator, the most common type in service today (Table 4).

The equal wall stator has evolved and now demonstrates improved pump capability. It is manufactured using a cast outer metallic shell containing the same lead thread as the molded inside diameter will have. Thus, the thickness of the elastomer lining is uniform throughout the length of the stator. While the cast stator is more expensive than pipe, it requires less volume of elastomer. If the elastomer is costly (e.g., fluorocarbon), the total cost of the stator may be lower, reducing both initial pump cost and replacement stator costs.

As the rotor turns inside an unequal wall stator, the work done on the elastomer (flexing of the elastomer) varies between where it is thick versus thin. This requires significantly more starting and running torque than an equal wall stator pump. The nonuniform flexing of the unequal wall stator liner generates nonuniform frictional heat within the elastomer with a nonuniform heat dissipation capability that limits its pressure per stage rating. The uniform flexing of the elastomer in an equal wall stator allows a pressure rating per stage in the order of twice as high as an unequal wall stator design of the same number of stages. At elevated pressure rating, abrasive solids content should be kept minimal for reasonably equal wall stator life. See Fig. 119.

A 1:2 geometry in a two-stage pump or a 2:3 geometry single-stage pump will meet the specified pumping requirements. More stages reduce the pressure rise per stage and will significantly extend pump life, especially in abrasive services.

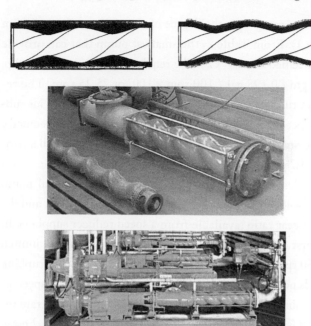

FIGURE 119
Standard (left) and "equal wall" (right) stator designs and comparison of profiles.

Example 2: Pump Sizing

Fluid: Polymer

Flow Rate: 5 gpm

Differential Pressure: 150 psid

Maximum Particle Size: None; liquid is clean, viscosity low
   (100 cP)

First, categorize the degree of abrasiveness ("none" from Table 1). From Table 2, the smallest pump that can provide the specified flow is "C." The required speed, neglecting slip, is 5 gpm / 0.198 gal / 100 rpm × 100, or 25.25 rpm. The resultant rubbing speed is 25.25 × 0.42, or 10.6 ft/sec. This rubbing speed is midrange of that recommended. A 1:2 geometry single-stage pump will meet the requirements, as will a two-stage 1:2 geometry unequal wall pump.

Unless a variable speed drive is used, the actual pump speed will depend upon the loaded speed of the driver and the actual gear ratio (available from a speed reducer supplier). In the first example, the actual available ratios might be limited to 380 rpm or 405 rpm, in which case flow rates and rubbing speeds need to be recalculated at the actual expected speed to determine pump performance. Obviously, if the flow rate required is critical, the higher-speed reducer should be chosen. If the required flow rate is not critical, the slightly lower ratio will produce slightly less flow and draw slightly less power. Note that in the PC pump industry, the standard direction of rotation is counterclockwise, facing the pump shaft. Most PC pumps are driven through speed reduction mechanisms that reverse the input drive direction of rotation.

Due to the interference fit of the rotor to the stator, the necessary pump starting torque can exceed the pump running torque requirement (Fig. 120). Both must be calculated, and the higher of the two is used to size the driver. The calculation methodology is often proprietary to the PC pump manufacturer. Allowance for speed reduction device inefficiency must also be factored into the final power rating of the drive needed.

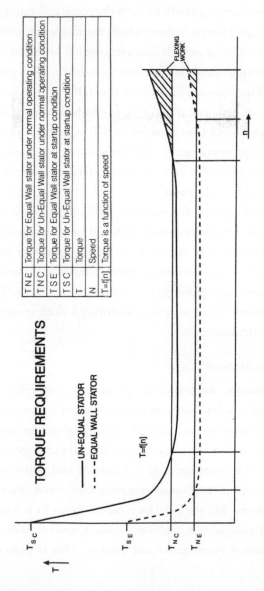

FIGURE 120 *Starting torque relationship (standard design and the equal-wall stator).*

Due to thermal growth of both the rotor and stator, for liquid pumping temperatures outside the temperature range of ~ 50–115°F, rotor outside diameters are left oversized (cold service) or undersized (hot service) to maintain the required amount of interference fit between them. If a rotor is undersized for high-temperature operation, standard factory testing will not achieve the same flow rate as will be pumped when in service at high temperature. The ambient test will produce a lower flow rate due to little or no interference between the new rotor and its new stator.

Some PC pumps are available with adjustable stators. These stators are axially slotted on the OD and equipped with clamping bands along their length. As the clamps are tightened, the degree of rotor-to-stator interference can be controlled under varying pumping temperature conditions. They can also compensate for wear by restoring a worn stator to a like-new interference fit.

Metallic Stators

Metallic stators, also known as rigid stators, can be supplied when requirements, operating conditions, and economies are aligned. Metallic stators can handle much higher pressure per stage than elastomeric stators, up to 500 psi per stage. They are used with pumpage over ~ 5,000 cp and allow much shorter pumps for high-pressure service than would otherwise be possible for PC pumps. The rotor-to-stator fit is a clearance fit. Consequently, removal, cleaning, and reinstallation of the rotor is much easier and quicker. This is especially advantageous where cleaning every shift is a requirement such as in food plants. Products pumped most frequently in the food industry include meat emulsions, cookie fillings, cake and cookie icings, glucose, glues, pastes, hot grease, and molasses, as well as paint, hot resins, varnishes, and similar high-viscosity materials.

Metallic stators are available in various stainless steels as well as tool steels. Since there is essentially no rotor-to-stator contact, product contamination from elastomer wear particles is eliminated. Metallic stator PC pumps can handle higher temperatures as well, to 500°F with drive-end modifications. They

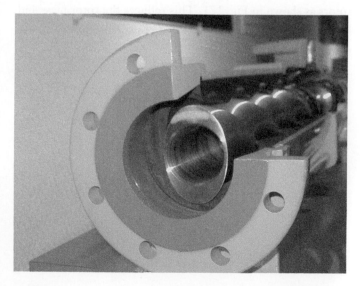

FIGURE 121
Hollow rotor design.

are more abrasion-resistant for the same pressure per stage as a nonmetallic stator, and can have useful life up to ten times that of an elastomeric stator. They have a broader chemical compatibility than is possible with most elastomers. When used as a high-pressure, low-number-of-stages pump, the initial pump cost may be similar to an elastomeric design with a higher number of stages. Generally, maximum speed for pumps fitted with metallic stators is 400 rpm and a maximum particle size of 200 microns. Using a fewer stage metallic stator (and corresponding rotor), there is less viscous drag on the rotor and the reduced friction improves pump operating efficiency.

There are other specialty designs available, such as hollow rotors as shown in Fig. 121. This reduces the rotor mass and helps reduce unbalance forces and rotor vibrations, and extends life.

Chapter 10

Progressing Cavity Pump Startups

Many PC pump startups are the culmination of months, if not years, of work designing the process, machine or system, specifying components, instrumentation, protective devices, reviewing and qualifying suppliers, etc. It is also the most vulnerable time for any pump. This section describes cautions, reviews, and inspections that should be conducted before startup to help ensure that all of those many gremlins of pumping systems are found out and addressed in time.

Thoroughly read the technical manuals and instructions from the pump, driver, and all auxiliary equipment suppliers to uncover requirements that may be specific to their equipment design. This is the easiest method to protect the system but is overlooked more often than not.

Pipe and Valves

Piping and valving installation should probably be considered first. Be sure all required valves have been installed. Verify that none are installed backwards. An absent or reverse-mounted check valve, foot valve, or relief valve can cause some very serious damage. Piping should have been inspected during fabrication to ensure that weld bead, weld rod, scale, etc. have been completely removed. Such hard particles can cause pump damage should they lodge in the wrong pump clearance. Temporary or permanent pump inlet strainers should be considered if they are not present, and they should start in a clean condition so that accumulation of dirt can be monitored. Obviously, if there is solids content in the liquid, strainers may not be necessary or desirable, but strainers offer inexpensive protection for the pump and valves in clean services.

The piping system should be pressure tested. Avoid imposing pressures in excess of any system component's design limits. Many pumps can withstand discharge pressure only on their discharge side. Inlet piping systems are frequently suitable only for low pressure. The pressure test medium should be compatible with the components and system being tested. A low pressure (e.g., 15 psig) compressed air test may be adequate to find missing flange gaskets or other obvious leak sources.

Check and tighten all flange bolts to their specified torque. The pump inlet and discharge piping should begin approximately 20 feet (6 meters) away from the pump to minimize pipe strain on the pump. Piping should be independently supported—pumps do not make very good pipe anchors! When pipe flanges are unbolted from the pump, the flange bolts should be able to be installed or removed without forcing the piping into position. There should be a flange-to-pump gap not exceeding the greater of twice the flange gasket thickness, or $\frac{1}{16}$ in. (3 mm). If either is not true, rework the piping until the gap meets this specification.

Positive displacement pumps will normally have a system pressure relief valve installed from the discharge piping to either the source of the pumped liquid, such as a supply tank, or to the pump inlet piping (a less desirable point due to the potential for temperature buildup during relief valve operation). This valve will normally be set slightly higher than the maximum anticipated normal system operating pressure. If possible, verify that it has been properly set. If this cannot be verified, consider setting the relief valve to a very low pressure and adjusting it upward after pump startup. Consult the relief valve vendor's technical data to be sure valve adjustment is done in the correct direction to lower the pressure. When pumping high solids content liquids, relief valves may not be functional and a pressure switch to shut down the pump drive may be a better choice to protect the pump, drive, and system from the adverse effects of excessive discharge pressure. Rupture (blow-out) disks are also a viable means of over-pressure protection when handling high solids content liquids. These are one-time-use thin metal disks designed to rupture at a specific pressure. The downstream side of the disk must be piped to a location that can accept flow after disk rupture until the pump can be stopped.

Ideally, the entire piping and valve system will be thoroughly flushed to remove all dirt and fabrication debris. This is normally done using a flush pump, not the normal system pump. As applicable, strainers and/or filters are installed at appropriate locations; their dirt accumulation is monitored until they show no accumulation for a period of 24 hours. Flushing usually uses light, fairly hot (150°F) liquid delivered at flow rates higher than system design. The higher flow rates cause higher liquid velocities within the piping system and are more likely to dislodge debris. Some systems use vibration equipment to mechanically shake the piping, again to improve dislodging dirt. Very extensive piping systems have been known to still show debris accumulation after 30 days of flushing.

Before final startup, be sure valves are open or closed as required. Pump inlet and discharge valves are normally left fully open. Manual pump bypass valves are also normally left open on startup. An air bleed valve in the discharge piping at a high point near the pump will significantly improve the pump's ability to self-prime. The valve is left open during startup until liquid flows, and then it is shut. Be sure to know where this flow will be directed to avoid inadvertent spillage or discharge to the atmosphere. Steam turbine valving is very important. Turbine startup procedures should be thoroughly reviewed as there are personal injury issues associated with this equipment if started or operated improperly.

Foundation, Alignment, and Rotation

If horizontal pumps are used, be sure the foundation is level, that hold-down bolts are tight, and that grouting, if used, has completely filled the base plate (no hollows or voids) and has cured.

If the pump will be handling liquid above ~ 150°F, a steam turbine is used as the driver and an estimate of the hot machine's centerline growth in height must be made. Shaft-to-shaft alignment (cold) should incorporate a deliberate compensating offset so that the alignment is more nearly correct when the equipment is up to operating temperature. Coefficients of thermal expansion for common pump case materials are shown in Table 5.

Table 5
Thermal Growth Coefficients (×10⁻⁶)

Material	*inches/inch/°F*
Cast iron	6.0
Ductile iron	6.6
Cast steel	6.5
316 stainless steel	9.4

The coefficient is applied to the centerline height of the shafts and the *difference* in temperature between that at which the unit was aligned and the temperature of expected operation. The cold machine should be shimmed high by the calculated distance shown in Table 5.

The purpose of any shaft aligning procedure is to align the centers of the machine shafts with each other, *not* to align the flexible coupling hubs. At operating temperature, alignment should be within 0.003 in. of total indicator reading (TIR), both angular and parallel. Consult a good aligning procedure to achieve or verify this degree of precision. The fact that the coupling may be rated to a much greater misalignment capability has nothing to do with the shaft-to-shaft alignment of the equipment. Survival and longevity of the machinery, *not* the coupling, are the objectives. If hot pumps and/or drivers are used, after they are at nominal operating temperature long enough for thermal growth to have stabilized (probably one hour or more), shut down the equipment and verify that alignment is within proscribed limits.

Never rely upon the alignment that was produced where the pump and drivetrain were assembled. Transportation, lifting, and handling, as well as foundation irregularities, will impact alignment always in an undesirable direction. Final alignment should be established as nearly the last step before actually starting the pump. If the equipment is to be dowelled in place, do so to the pump *only* after several hours, if not days, of good operation and alignment checks.

The use of resilient mounts is sometimes desired to reduce vibration being transmitted into the underlying foundation. Such mounts must not be installed beneath the pump or driver, but between the pump/driver base plate or bracket and the foundation. The pump and driver must be rigidly aligned, not resiliently aligned, as the resilient mounts will not maintain adequate alignment under torsional reactions from the transmitted torque.

The direction of rotation is critical for most equipment. It is usually indicated by arrow nameplates. Remember that some gearing will reverse rotation from the input shaft to the output shaft. Most engines and turbines must be purchased for a specific direction of rotation, and this is also true of most pumps. Standard AC electric motors are frequently bidirectional; their direction of rotation will depend upon how the power cables are connected. It is normally not possible to predict their direction of rotation beforehand. It is recommended that the flexible coupling at the motor shaft be disconnected and the motor momentarily energized (jogged on, then immediately off) to see if its rotation is correct for the rest of the driven equipment. If it is not, two of the electric power cables will need to have their connections reversed. Verify correct rotation after reversing, if necessary, before re-engaging the flexible coupling.

Lubrication

Most rotating machinery has some form of lubrication for its bearing systems. It may be as simple as a permanently grease-packed, sealed ball bearing or as complicated as a separate lubricating oil-pump system complete with cooler, filter, instrumentation, etc. Be sure to verify that any required lubrication has been addressed. Equipment having been in storage may require draining and addition of fresh lubricant or even flushing out of preservatives before fresh lubricant is added. Any gearing present (reduction drive gears, etc.) should be reviewed for the presence of the correct type and quantity of lubricant. Constant level oilers should be filled to their mark with clean, fresh lubricant of the correct type. Some flexible couplings are grease-lubricated and should also be checked. Most electric motors will have grease-lubricated antifriction bearings that should be checked as well. PC pump eccentric joints, if lubricated, are normally filled with grease or oil at the pump manufacturer's plant.

Almost all rotary pumps should be able to be turned over by hand. They should generally turn over smoothly, with no catches or uneven rubbing. Large pumps will need a helper bar but should not be difficult to turn. If not, consult the pump vendor. Partial disassembly may be advisable to determine the cause of such difficulty (foreign material, rust, etc.) before starting.

Startup Spares

With care and planning, startups will generally go smoothly, without significant problems. However, it is prudent to have key spare parts on hand in the event they are needed quickly for correction after some unanticipated problem, minor damage, or need to disassemble a piece of equipment for inspection. For PC pumps, this would normally be a set of shaft seals, gaskets, O-rings, and bearings, frequently available as a minor repair kit. For other rotating equipment, spare bearings, grease and oil seals, gaskets, etc. should be on hand to avoid delay in the pump startup. More extensive spares will depend on availability from the vendor, criticality of pump operation, plant practice and, perhaps, other issues specific to the installation. If the startup goes well and the spares are not consumed, they are very appropriate to be held on hand for future routine inspections and service.

Resources

Be sure electric power, steam, cooling water, hot oil, instrumentation power or air, or any other auxiliary resources are available and ready before startup. Be sure adequate pressure and temperature gages are in place so observations can be made during startup. Without them, you are working blind. Speed indicators (tachometers) may also be needed if the drive is not a fixed-speed one, such as an AC electric motor.

Last-Minute Startup Thoughts

It is good practice to initially fill the pump and as much of the inlet piping system as possible with the liquid to be pumped. This will assist in priming and reduce the risk of pump damage during an otherwise dry start. A rotary pump will prime more quickly if internal pumping elements are at least wetted. Priming is nothing more than pumping air from the inlet system to the discharge system. The ability of a rotary positive displacement pump to act as an air compressor is very much related to having some internal liquid present.

Pump shaft seals, especially mechanical seals, should never be operated dry. Immediate, or at best premature, seal failure is the inevitable result. Again, filling the pump with the liquid to be pumped and hand-rotating the pump a few times help ensure that liquid is present at the shaft sealing mechanism to carry away frictional heat during startup. If the particular pump has a seal chamber access plug, remove the plug, fill the chamber with liquid, and reinstall the plug.

Have phone numbers of key vendor service departments, the fire brigade, and medical emergency services on hand in the event they are needed. Depending on the liquid pumped, there might be pollution and fire hazards present.

Ensure that there is an adequate supply of liquid in the pump inlet system (no half-empty supply oil tanks or the like). It is also prudent to confirm where the pump discharge flow will be going to be sure the discharge system is ready.

Loud or erratic noise at startup is an indication of cavitation (inadequate pump inlet pressure) or air being drawn into the pump inlet system. It is frequently accompanied by increases in or excessive vibration. If this is mild, troubleshoot the cause. If it is severe, shut down the pump and find the source of the problem.

Use the Rotary Pump Startup Checklist (see page 201) or a similar control to help ensure that all contingencies have been addressed. Each pumping system has unique features and requirements, some of which may interact with each other or with other aspects of the overall plant operation. In addition, no allowance has been made here for regulatory requirements, specialized industry or company guidelines, or the like. Where operational values are recommended, they are intended for use in the absence of vendor or specifically engineered information. Always use the more stringent of either the recommendations herein or the vendor's or engineer's guidelines.

Rotary Pump Startup Checklist

*Project:*_____ *Location:*_____ *Unit No.:*_____
*Tag No.:*_____

1. Piping

☐ Clean
☐ Bolts tight
☐ Strain removed
☐ Gaskets in place
☐ Pressure tested
☐ Flushed
☐ Other_____

2. Valves

☐ Not backwards
☐ Clean
☐ Bolts tight
☐ Gaskets in place
☐ Correct position (open/closed)
☐ Relief valve set pressure
☐ Other_____

3. Foundation

☐ Level
☐ Solid (no voids)
☐ Bolts tight
☐ Other_____

4. Alignment

☐ Angular (cold & hot)_____

☐ Parallel (cold & hot)_____

☐ Other_____

5. Rotation

☐ Verified (CW or CCW)

☐ Other_____

6. Lubrication

☐ Pump

☐ Driver

☐ Gear

☐ Other_____

7. Spares Available

☐ Pump

☐ Driver

☐ Gear

☐ Other_____

8. Resources Available

☐ Electric

☐ Steam

☐ Cooling water

☐ Hot oil

☐ Auxiliaries

☐ Gages in place

☐ Other_____

9. Last-Minute

☐ Pump filled with liquid
☐ Shaft seals wetted
☐ Key contact phone numbers
☐ Inlet liquid supply
☐ Discharge system ready
☐ Air bleed valve open
☐ Pump hand-rotated
☐ Other_____

10. Company- or Industry-Specific

☐ _____
☐ _____
☐ _____
☐ _____

Chapter 11

PROGRESSING CAVITY PUMP
OVERHAUL GUIDE

A progressing cavity pump overhaul can be handled on a professional basis with clearly understood expectations and communications, regardless of whether the repair is to be done at the pump manufacturer's facility, in-house, or by a third party (on- or off-site).

Having the pump manufacturer (original equipment manufacturer) perform the overhaul has some obvious advantages, not least of which is their access to original component design, dimensional, and tolerance data, material specifications, and their experience with the pump. They also usually have complete testing facilities. New pump warranties are sometimes available from the original pump manufacturer, as well as upgrades and modernizations.

Third-party repair is frequently less costly, at least on a first-cost basis, but rarely includes a warranty, almost never includes any meaningful testing, and can sacrifice reliability for expediency.

In-house overhaul, facilities and personnel permitting, is a good alternative if some in-house expertise or experience is available. Unfortunately, with the many company "right-sizing" initiatives occurring of late, maintenance personnel are frequently the first to go, with resultant loss of in-house knowledge as well as historical perspectives that frequently led to very quick analyses and fixes.

If there are company- or industry-specific or unique requirements to be imposed on the overhaul, such as sanitary specifications or testing liquids compatibility, be sure they are spelled out in writing before proceeding. If troubleshooting, failure analysis, or other services are expected or needed, they too should be made clear before the overhaul commences.

Regardless of the overhaul facility location, the basic steps to be followed are similar. Pumps for overhaul should be delivered to the overhaul facility as clean as possible and include a material safety data sheet (MSDS) for any liquid residue in the pump(s). The basic overhaul steps include (as applicable):

1. Received condition report (external damage, missing parts, extraneous parts, etc.)

2. Review of MSDS to ensure proper handling of residue liquid in pump

3. Disassembly of the pumps

4. Thorough cleaning of pump component parts for inspection

5. Inspection of each part to determine if the part should be

- *Scrapped* (always for elastomers and gaskets, usually for antifriction bearings and frequently for mechanical seals unless sufficiently valuable to have them overhauled themselves)

- *Reworked* (remachined, plated, scraped, polished, weld build-up, etc.)

- *Reused* as-is

Rotary Pump Inspection Report

Owner _____   Model No. ___
Serial No. _____   MSDS ___Yes ___No
Received Condition _____

Part I.D.	Qty	Part No.	Part Name	Recommended Disposition- Check one			Price
				Scrap	Reuse	Rework & Comments	$

By _____   Subtotal $
Date _____   Assembly/Test
Comments _____   Other_____
                                    Other_____
                                    Total _____ $

6. A written inspection report relating the above component condition/disposition recommendations should be produced. The report should include pricing for the recommended overhaul, noting what is or is not included (painting, testing, special preservation, etc.).

If a warranty claim or troubleshooting effort is to be made, be sure to keep *all* parts, regardless of condition, so a more complete evaluation can be made.

A comparison to the cost of a new pump may be appropriate if such is still manufactured. Many off-the-shelf pumps are manufactured in large lot sizes and sell for about the same price as an overhaul handled on a one-time basis.

The party paying for the overhaul should review these recommendations and, depending on cost, time, criticality, etc., either challenge the observations if any appear questionable or authorize the overhaul to commence. Note that time will frequently determine the reuse of questionable or worn parts. The overhaul should specifically state that new pump performance will be achieved (or something less) due to economics or time. A loading or unloading pump flow rate may not be very critical to day-to-day operations, while pump capacity may be critical for gas sealing, machinery lubrication, or process applications. If something less than new pump performance is expected, then the minimum flow rate should be agreed upon before the overhaul proceeds.

If weld buildup or repair is to be conducted, be sure that the weld procedures and welders have been adequately qualified. Plating and spray buildup processes are frequently technique-sensitive and may need some nondestructive testing after application to confirm whether the objectives were met.

Not all pumps are tested after repair. If this is a requirement, specify so in the ordering document. Testing ranges from turning the pump shaft by hand, to hydrostatic and spin tests, to full qualification testing as would be conducted on a new pump. If new pump flow rate is expected after the repair, again, so specify. If the overhauled pump will be installed and operated upon its return, no special preservation or packaging may be needed. On the other hand, if the pump will go into storage, internal preservation and sturdy crating should be specified. Application of a paint should be included if the pump exterior is of cast iron, steel, or other material susceptible to rusting.

Conclusion

In this book we have looked at the geometry of the progressing cavity family of machines, which included *surface pumps (PCs)*, *downhole pumps (DHPs)*, and *mudmotors (DHMs)*. Using basic geometry and minimum mathematics, the complexity and mystery of the three-dimensional shapes of these machines has been unveiled and better understood. Practical aspects, such as troubleshooting techniques, application guidelines, and installation examples also have been included. A reader, student, practicing engineer, or the equipment user, should find this book a good reference for his or her work with PC surface-mounted and downhole pumps, and mudmotors. It could also help the reader to be able to better understand, select, and apply these types of pumps for actual *applications* in chemical and petrochemical plants, pulp and paper mills, waste treatment facilities, and many other facilities. Proper application, utilizing the capabilities and versatility of these units, as well as appreciation of their limitations, would help improve equipment reliability and life, provide a better economic evaluation, and enable proper selection of the required unit size, speed, flow, torque, and horsepower requirements based on real operating conditions. Pump selection and sizing has also been presented.

Comments, suggestions, corrections, or additions are welcome to improve readability, enhance accuracy, and to make this publication even more useful for the readers of future editions.

Please send your comments to:

Dr. Lev Nelik, P.E., APICS
Pumping Machinery, LLC
Atlanta, GA
Tel.: 770-310-0866
DrPump@PumpingMachinery.com
www.PumpingMachinery.com

Index

REFERENCES

1. Hydraulic Institute Pump Standards, ANSI/HI 1.1–1.5, 1994.

2. Dresser Company Publication Catalog, *Security Downhole Motor Services Operations Handbook*, Houston, TX, November 1992.

3. H. Cholet, "Progressing Cavity Pumps," *Editions Technip*, Paris, 1997.

4. L. Nelik, "Centrifugal and Rotary Pumps: Fundamentals, Comparisons and Applications." Short Course, ASME Professional Development Program, Pittsburgh, PA, June 1998.

5. L. Nelik, "Positive Displacement Pumps," Short Course, 15th International Pump Users Symposium, Texas A&M University, Houston, TX, March 2000.

6. L. Nelik, "Pumps," in *Kirk-Othmer Encyclopedia of Chemical Technology*, 4th Ed., Volume 20. New York: John Wiley & Sons, Inc., 1996.

7. R. Samuel and K. Saveth, *Progressing Cavity Pump (PCP): New Performance Equations for Optimal Design*. Richardson, TX: Society of Petroleum Engineers, 1996.

8. M. Delpassand, "Progressing Cavity Pump Design Optimization for Abrasive Applications," *Artificial Lift*, July 1997.

9. G. Vetter and W. Wirth, "Understanding Progressing Cavity Pumps Characteristics and Avoid Abrasive Wear," paper presented at the Texas A&M University 12th International Pump Users Symposium, March 1995.

10. D. Baldenko, F. Baldenko, and A. Shmidt, "Screw Downhole Motors: New Designs and Control Methods," *Downhole Drilling* (in Russian), 1997.

Printed and bound by CPI Group (UK) Ltd, Croydon, CR0 4YY

03/10/2024

01040432-0001